AGRICULTURAL CHEMISTRY

Edited by
Levitah Castil Mapatac

Delve Publishing

Agricultural Chemistry

Edited by Levitah Castil Mapatac

Edition 2018

Delve Publishing

Published by

Delve Publishing LLC

708 3rd Avenue, 6th Floor

New York

NY 10801

United States of America

www.delvepublishing.com

ISBN: 978-1-68095-721-1 (Hardback)

Printed in United States of America

Library of Congress Control Number: 2016920401

About the Editor

Levitah Castil Mapatac

Levitah C. Mapatac is an Associate Professor, researcher, writer and a PhD holder from the University in Science and Technology of Southern Philippines. Her interest varies from different chemistry concepts and applications and currently doing research in traditional medicines, water waste pollution and environmental chemistry in the aspect of contamination pathways, profilling and rehabilatation of mining waste.

Table of Contents

2 ANALYSIS OF WATER CHEMISTRY

3 SOIL-FORMING PROCESSES AND PROPERTIES

4 SOIL: CHEMICAL AND PHYSICAL CHARACTERS

5 SOIL BACTERIA

List of Abbreviations

ACP	Acyl Carrier Protein
AEC	Anion Exchange Capacity
CEC	Cation Exchange Capacity .
CWA	Chemical Warfare Agents
CCD	Colony Collapse Disorder
CHP	Combined Heat And Power
DME	Dimethyl Ether
DOM	Dissolved Organic Matter
EAI	Electronic Associates, Inc.
EI	Electron Ionization
ETBE	Ethyl Ter-Butyl Ether,
EFOA	European Fuel Oxygenates Association
FAMEs	Fatty Acid Methyl (or Ethyl) Esters
FOM	Fresh Organic Matter
HTU	Hydrothermal Upgrading
IGR	Insect Growth Regulator
ICSB	International Committee on Systematic Bacteriology
MTBE	Methyl Ttertiary-Butyl Ether
MAL	Mobile Analytical Laboratories
MRM	Multiple Reaction Monitoring
NCI	Negative Chemical Ionization
NMR	Nuclear Magnetic Resonance

POM	Particulate Organic Matter
PCI	Positive Chemical Ionization
SIM	Selected Ion Monitoring,
SRM	Selected Reaction Monitoring
SOM	Soil Organic Matter
SOM	Soil Organic Matter
SPME	Solid-Phase Micro Extraction
SMB	Supersonic Molecular Beam
WHO	World Health Organization

Preface

Agricultural chemistry is the study of both chemistry and biochemistry which are important in agricultural production, the processing of raw products into foods and beverages, and in environmental monitoring and remediation. Agricultural chemistry often aims at preserving or increasing the fertility of soil, maintaining or improving the agricultural yield, and improving the quality of the crop. Carbon inputs from photosynthetic organisms ultimately provide the fuel for many soil organisms to grow and reproduce. Soil organisms, in turn, promote organic carbon degradation and catalyze the release of nutrients required for plant growth. The stability and productivity of agricultural ecosystems rely on efficient functioning of these and other processes, whereby carbon and nutrients such as nitrogen and phosphorus are recycled. Human-induced perturbations to the system, such as those that occur with pesticide or fertilizer application, alter ecosystem processes, sometimes with negative environmental consequences. Soil is the primary medium in which biological activity and chemical reactions occur. It is a three- phase system consisting of solid, liquid, and gas. Approximately 50 percent of the volume in a typical agricultural soil is solid material classified chemically as either organic or inorganic compounds. Organic materials usually constitute 1 to 5 percent of the weight of the solid phase. The remainder of the soil volume is pore space that is either filled with gases such as CO2 and O2, or water. A wide variety of organisms live in soil, including microorganisms not visible to the naked eye such as bacteria, fungi, protozoa, some algae, and viruses. Bacteria are present in the largest numbers, but fungi produce more biomass per unit weight of soil than any other group of microorganisms. Much of agricultural chemistry as it relates to nutrient cycles, pesticide transformation, plant growth, and organic matter degradation involves the participation of microorganisms. Microorganisms produce both intracellular and extracellular enzymes that increase reaction rates, oxidize and reduce organic and inorganic compounds, and synthesize organic molecules that modify soil chemical and physical properties. Agricultural chemicals have an important role in land management and the production of food and fiber products, protecting plants from damage caused by insect pests, weeds or disease. In most cases, agricultural chemicals are referred to as 'pesticides' and include herbicides, fungicides, insecticides and pest animal baits.

Content Coverage

Chapter One presents the introduction of organic matter composed of the remains of animals and plants and the waste products the organisms leave in the environment.

Chapter Two focuses on chemical water analysis carried out on water used in agricultural processes. Water chemistry analysis is often the groundwork of studies of water quality, pollution, hydrology and geothermal waters.

Chapter Three outlines the soil-forming processes and properties. As soil resources serve as a basis for food security, the international community advocates its sustainable and responsible use through different types of soil governance.

Chapter Four highlights about the study of the chemical characteristics of soil. Soil chemistry is affected by mineral composition, organic matter and environmental factors.

Chapter Five focuses on the role of bacteria in soil. Bacteria are single celled microbes. The cell structure is simpler than that of other organisms as there is no nucleus or membrane bound organelles.

Chapter Six emphasizes on the organic compounds, which is virtually any chemical compound that contains carbon, although a consensus definition remains elusive and likely arbitrary. Organic compounds are rare terrestrially, but of central importance because all known life is based on organic compounds.

Chapter Seven focuses on the various chemical products used in agriculture. In most cases, agrichemical refers to the broad range of pesticides, including insecticides, herbicides, fungicides and nematicides. It may also include synthetic fertilizers, hormones and other chemical growth agents, and concentrated stores of raw animal manure.

Chapter Eight presents the study of fertilizers and nitrate leaching. Nitrate is one of the facts of life. It is essential for the growth of many plant species, including most of those we eat, but it becomes a problem if it gets into water in which it is not wanted.

CHAPTER 1

Organic Matter: An Introduction

Chapter Contents

- **Introduction**
- **Formation**
- **Natural Ecosystem Functions**
- **Source Cycle**
- **Soil Organic Matter**
- **Decomposition**
- **Organic Chemistry**
- **Organic Matter in Water (Aquatic)**
- **Vitalism**
- **Biomass**
- **Biofuel**
- **Humus**
- **Detritus**

Organic matter is composed of the remains of animals and plants and the waste products the organisms leave in the environment. Unlike organic material, organic matter has decomposed to the point that it is resistant to decomposition. This matter is decomposed into a humus that has been converted by microorganisms.

INTRODUCTION

Organic matter, organic material, or natural organic matter (NOM) refers to the large pool of carbon-based compounds found within natural and engineered, terrestrial and aquatic environments. It is matter composed of organic compounds that has come from the remains of organisms such as plants and animals and their waste products in the environment. Organic molecules can also be made by chemical reactions that don't involve life. Basic structures are created from cellulose, tannin, cutin, and lignin, along with other various proteins, lipids, and carbohydrates. Organic matter is very important in the movement of nutrients in the environment and plays a role in water retention on the surface of the planet.

FORMATION

Living organisms are composed of organic compounds. In life they secrete or excrete organic materials into their environment, shed body parts such as leaves and roots and after the organism dies, its body is broken down by bacterial and fungal action. Larger molecules of organic matter can be formed from the polymerization of different parts of already broken down matter. The composition of natural organic matter depends on its origin, transformation mode, age, and existing environment, thus its bio-physico-chemical functions vary with different environments.

NATURAL ECOSYSTEM FUNCTIONS

Organic matter is present throughout the ecosystem. After degrading and reacting, it can move into soil and mainstream water via waterflow. Organic matter provides nutrition to living organisms. Organic matter acts as a buffer in aqueous solution to maintain a neutral pH in the environment. The buffer acting component has been proposed to be relevant for neutralizing acid rain.

SOURCE CYCLE

A majority of organic matter not already in the soil comes from groundwater. When the groundwater saturates the soil or sediment around it, organic matter can freely move between the phases. Groundwater has its own sources of natural organic matter also:

- “organic matter deposits, such as kerogen and coal
- soil and sediment organic matter
- organic matter infiltrating into the subsurface from rivers, lakes, and marine systems

Note that one source of groundwater organic matter is soil organic matter and sedimentary organic matter. The major method of movement into soil is from groundwater, but organic matter from soil moves into groundwater as well. Most of the organic matter in lakes, rivers, and surface water areas comes from deteriorated material in the water and surrounding shores. However, organic matter can pass into or out of water to soil and sediment in the same respect as with the soil.

Importance of the Cycle

Organic matter can migrate through soil, sediment, and water. This movement enables a cycle. Organisms decompose into organic matter, which can then be transported and recycled. Not all biomass migrates, some is rather stationary, turning over only over the course of millions of years.Janna Jajaijsnsjs Nani

SOIL ORGANIC MATTER

Soil organic matter (SOM) is the organic matter component of soil, consisting of plant and animal residues at various stages of decomposition, cells and tissues of soil organisms, and substances synthesized by soil organisms. SOM exerts numerous positive effects on soil physical and chemical properties, as well as the soil's capacity to provide regulatory ecosystem services. Particularly, the presence of SOM is regarded as being critical for soil function and soil quality.

The positive impacts of SOM result from a number of complex, interactive edaphic factors; a non-exhaustive list of SOM's effects on soil functioning includes improvements related to soil structure, aggregation, water retention, soil biodiversity, absorption and retention of pollutants, buffering capacity, and the cycling and storage of plant nutrients. SOM increases soil fertility by providing cation exchange sites and acting as reserve of plant nutrients, especially nitrogen (N), phosphorus (P), and sulfur (S), along with micronutrients, which are slowly released upon SOM mineralization. As such, there is a significant correlation between SOM content and soil fertility.

SOM also acts as a major sink and source of soil carbon (C). Although the C content of SOM is known to vary considerably, SOM is typically estimated to contain 58% C, and the terms 'soil organic carbon' (SOC) and SOM are often used interchangeably, with measured SOC content often serving as a proxy for SOM. Soil represents one of the largest C sinks on the planet and plays a major role in the global carbon cycle. Therefore, SOM/SOC dynamics and the capacity of soils to provide the ecosystem service of carbon sequestration through SOM management have received considerable attention in recent years.

The concentration of SOM in soils generally ranges from 1% to 6% of the total topsoil mass for most upland soils. Soils whose upper horizons consist of less than 1% organic matter are mostly limited to desert areas, while the SOM content of soils in low-lying, wet areas can be as high as 90%. Soils containing 12-18% SOC are generally classified as organic soils.

It can be divided into three general pools: living biomass of microorganisms, fresh and partially decomposed residues, and humus: the well-decomposed organic material. Surface plant litter is generally not included as part of soil organic matter.

Sources of Soil Organic Matter

The primary source of organic matter contained in soil is vegetal. In forest or prairies, as well as agricultural fields, dead plants are transformed by different kinds of living organisms. This process involves several steps, the first being mostly mechanical, and becoming more chemical as it progresses. The small living beings that work on that decomposition chain are themselves part of the soil organic matter, and form a food web of organisms that prey upon each other and are preyed upon.

There are also other animals that consume living vegetal material, whose residues are passed to the soil. The products from the living organisms metabolism are the secondary sources of soil organic matter that also includes the dead corpses of these organisms. Some animals, like earthworms, ants and centipedes contribute to both vertical and horizontal translocation of organic material.

Additional sources of soil organic matter include plant root exudates and charcoal.

Composition of plant residues

The water content of most plant residues is in the range of 60% to 90%. The dry matter consists of complex organic material composed mainly of carbon, oxygen and hydrogen. Although these three elements make up about 92% of the dry weight of the organic material in soils, there are other elements that are of great importance for the nutrition of plants. They include nitrogen, sulfur, phosphorus, potassium, calcium, magnesium and a range of micronutrients.

Organic compounds found in plant residues include:

- **Carbohydrates** are made up of carbon, hydrogen and oxygen, and range in complexity from rather simple *sugars* to the big molecules of *cellulose.*
- **Fats** consist of glycerids of fatty acids, like butyric, stearic, oleic. They are also made up of carbon, oxygen and hydrogen atoms.
- **Lignins** are complex compounds that form the older parts of wood, and consist also mainly of carbon, oxygen and hydrogen. They are resistant to decomposition.
- **Proteins** contain nitrogen in addition to carbon, hydrogen and oxygen, and also small amounts of sulfur, iron, and phosphorus.
- **Charcoal** is elemental carbon derived from incomplete combustion of organic matter. Charcoal is resistant to decomposition.

Decomposition

The vegetal residues in general are not water-soluble, and they are not usable by the plants. They constitute, nevertheless, the raw materials from which plant nutrients are derived. The decomposition is carried out as enzymatic biochemical processes by soil microorganisms, that obtain the necessary energy from the same residues, and produce the mineral compounds that are apt to be absorbed by plant roots. This process by which organic compounds are broken down and transformed into mineral (inorganic) compounds is also referred to as *mineralization.* A portion of organic material is not mineralized, but transformed into stable organic matter *humus.*

The break down of the organic compounds is done at very different rates, depending on their nature. The ranking, from fast to slow rates are as follow.

- Sugars, starches and simple proteins.
- Proteins
- Hemicelluloses
- Cellulose
- Lignins and fats.

The reactions that take place can be included in one of three groups:

- Enzymatic oxidation, that produces carbon dioxide, water and heat. It affects the bulk or major portion of the material.
- The essential elements, nitrogen, sulfur, phosphorus are liberated and mineralized by a series of specific reactions.
- Compounds that are resistant to microbial action are formed by modification of original compounds or by synthesis of new ones by microbes, creating humus.

The list of mineral end products are as follow:

Element	Mineral end products
Carbon	CO_2, CO_3^{2-}, HCO_3^-, CH_4, C
Nitrogen	NH_4^+, NO_2^-, NO_3^-, N_2 (gas)
Sulfur	S, H_2S, SO_3^{2-}, SO_4^{2-}, CS_2
Phosphorus	$H_2PO_4^-$, HPO_4^{2-}
Others	H_2O, O_2, H_2, H^+, OH^-, K^+, Ca^{2+}, Mg^{2+}, etc.

Humus

As vegetal material undergoes decomposition, some microbial resistant compounds are formed. These include modified lignins, oils, fats and waxes. Secondly, some new compounds are synthesized, like polysaccharides and polyuronids. These materials form the basis for humus. New reactions take place between these compounds and some proteins and other nitrogen containing products, incorporating thus nitrogen and avoiding its mineralization. Other nutrients are also protected in this way from mineralization.

Humic Substances Classification

There is a classification into three groups, based on solubility in acids and alkalis, and also related to stability.

- *Fulvic acid* is the group which contains the materials that have the lowest molecular weight, and are soluble in acids and alkali, and susceptible to microbial attack.
- *Humic acid* group contains the intermediate materials, with medium molecular weight, soluble in alkali, but insoluble in acid, and intermediate resistance to microbial attack.
- *Humin* is the generic name for the materials with highest molecular weight, that are darkest in color, insoluble in acid and alkali, and with the most resistance to microbial attack.

Role in Carbon Cycling

Soil plays a major role in the global carbon cycle, with the global soil carbon pool estimated at 2500 gigatons. This is 3.3 times the size of the atmospheric pool (750 gigatons) and 4.5 times the biotic pool (560 gigatons). The pool of organic carbon, which occurs primarily in the form of SOM, accounts roughly 1550 gigatons of the total global C pool, with the remainder accounted for by soil inorganic carbon (SIC). The pool of organic C exists in dynamic equilibrium between gains and losses; soil may therefore serve as either a sink or source of C, through sequestration or greenhouse gas emissions, respectively, depending on exogenous factors.

Factors Controlling Rates of Decomposition

- Environmental factors
  1. Aeration
  2. Temperature
  3. Soil Moisture
  4. Soil pH
- Quality of added residues
  1. Size of organic residues
  2. C/N of organic residues
- **Rate of decomposition of plant residues**, in order from fastest to slowest decomposition rates:
  1. Sugars, starches, simple proteins
  2. Hemicellulose
  3. Cellulose
  4. Fats, waxes, oils, resins
  5. Lignin, phenolic compounds

Priming Effect

The *priming effect* is characterized by intense changes in the natural process of soil organic matter (SOM) turnover, resulting from relatively moderate intervention with the soil. The phenomenon is generally caused by either pulsed or continuous changes to inputs of fresh organic matter (FOM). Priming effects usually result in an acceleration of mineralization due to a *trigger* such as the FOM inputs. The cause of this increase in decomposition has often been attributed to an increase in microbial activity resulting from higher energy and nutrient availability released from the FOM. After the input of FOM, specialized microorganisms are believed to grow quickly and only decompose this newly added organic matter. The turnover rate of SOM in these areas is at least one order of magnitude higher than the bulk soil.

Other soil treatments, besides organic matter inputs, which lead to this short-term change in turnover rates, include input of mineral fertilizer, exudation of organic substances by roots, mere mechanical treatment of soil or its drying and rewetting.

Priming effects can be either *positive* or *negative* depending on the reaction of the soil with the added substance. A positive priming effect results in the acceleration of mineralization while a negative priming effect results in immobilization, leading to N unavailability. Although most changes have been documented in C and N pools, the priming effect can also be found in phosphorus and sulfur, as well as other nutrients.

Löhnis was the first to discover the priming effect phenomenon in 1926 through his studies of green manure decomposition and its effects on legume plants in soil. He noticed that when adding fresh organic residues to the soil, it resulted in intensified mineralization by the humus N. It was not until 1953, though, that the term *priming effect* was given by Bingeman in his paper titled, *The effect of the addition of organic materials on the decomposition of an organic soil*. Several other terms had been used before *priming effect* was coined, including priming action, added nitrogen interaction (ANI), extra N and additional N. Despite these early contributions, the concept of the priming effect was widely disregarded until about the 1980s-1990s.

The priming effect has been found in many different studies and is regarded as a common occurrence, appearing in most plant soil systems.However, the mechanisms which lead to the priming effect are more complex then originally thought, and still remain generally misunderstood.

Although there is a lot of uncertainty surrounding the reason for the priming effect, a few *undisputed facts* have emerged from the collection of recent research:

- The priming effect can arise either instantaneously or very shortly (potentially days or weeks) after the addition of a substance is made to the soil.
- The priming effect is larger in soils that are rich in C and N as compared to those poor in these nutrients.
- Real priming effects have not been observed in sterile environments.
- The size of the priming effect increases as the amount of added treatment to the soil increases.

Recent findings suggest that the same priming effect mechanisms acting in soil systems may also be present in aquatic environments, which suggests a need for broader considerations of this phenomenon in the future.

DECOMPOSITION

One suitable definition of organic matter is biological material in the process of decaying or decomposing, such as humus. A closer look at the biological material in the process of decaying reveals so-called organic compounds (biological molecules) in the process of breaking up (disintegrating).

The main processes by which soil molecules disintegrates are by bacterial or fungal enzymatic catalysis. If bacteria or fungi were not present on Earth, the process of decomposition would have proceeded much slower.

ORGANIC CHEMISTRY

Measurements of organic matter generally measure only organic compounds or carbon, and so are only an approximation of the level of once-living or decomposed matter. Some definitions of organic matter likewise only consider organic matter to refer to only the carbon content, or organic compounds, and do not consider the origins or decomposition of the matter. In this sense, not all organic compounds are created by living organisms, and living organisms do not only leave behind organic material. A clam's shell, for example, while biotic, does not contain much organic carbon, so may not be considered organic matter in this sense. Conversely, urea is one of many organic compounds that can be synthesized without any biological activity.

Organic matter is heterogeneous and very complex. Generally, organic matter, in terms of weight, is:

- 45-55% carbon
- 35-45% oxygen
- 3-5% hydrogen
- 1-4% nitrogen

The molecular weights of these compounds can vary drastically, depending on if they repolymerize or not, from 200-20,000 amu. Up to one third of the carbon present is in aromatic compounds in which the carbon atoms form usually 6 membered rings. These rings are very stable due to resonance stabilization, so they are difficult to break down. The aromatic rings are also susceptible to electrophilic and nucleophilic attack from other electron-donating or electron-accepting material, which explains the possible polymerization to create larger molecules of organic matter.

There are also reactions that occur with organic matter and other material in the soil to create compounds never seen before. Unfortunately, it is very difficult to characterize these because so little is known about natural organic matter in the first place. Research is currently being done to figure out more about these new compounds and how many of them are being formed.

ORGANIC MATTER IN WATER (AQUATIC)

Aquatic organic matter can be further divided into two subsections: dissolved organic matter (DOM) and particulate organic matter (POM). They are typically differentiated by that which can pass through a 0.45 micrometre filter (DOM), and that which cannot (POM).

Detection of Aquatic Organic Matter

Organic matter plays an important role in drinking water and wastewater treatment and recycling, natural aquatic ecosystems, aquaculture, and environmental rehabilitation. It is therefore important to have reliable methods of detection and characterisation, for both short- and long-term monitoring. A variety of analytical detection methods for organic matter have existed for up to decades, to describe and characterise organic matter. These include, but are not limited to: total and dissolved organic carbon, mass spectrometry, nuclear magnetic resonance (NMR) spectroscopy, infrared (IR) spectroscopy, UV-Visible spectroscopy, and fluorescence spectroscopy. Each of these methods has its own advantages and limitations.

Water Purification

The same capability of natural organic matter that helped with water retention in soil creates problems for current water purification methods. In water, organic matter can still bind to metal ions and minerals. These bound molecules are not necessarily stopped by the purification process, but do not cause harm to any humans, animals, or plants. However, because of the high level of reactivity of organic matter, by-products that do not contain nutrients can be made. These by-products can induce biofouling, which essentially clogs water filtration systems in water purification facilities, as the by-products are larger than membrane pore sizes. This clogging problem can be treated by chlorine disinfection (chlorination), which can break down residual material that clogs systems. However, chlorination can form disinfection by-products.

Potential Solutions

Water with organic matter can be disinfected with ozone-initiated radical reactions. The ozone (three oxygens) has very strong oxidation characteristics. It can form hydroxyl radicals (OH) when it decomposes, which will react with the organic matter to shut down the problem of biofouling.

VITALISM

The equation of "organic" with living organisms comes from the now-abandoned idea of vitalism that attributed a special force to life that alone could create organic substances. This idea was first questioned after the artificial synthesis of urea by Friedrich Wöhler in 1828.

BIOMASS

Sugarcane plantation in Brazil. Sugarcane bagasse is a type of biomass.

A cogeneration plant in Metz, France. The station uses waste wood biomass as an energy source, and provides electricity and heat for 30,000 dwellings.

Stump harvesting increases the recovery of biomass from a forest.

Biomass is organic matter derived from living, or recently living organisms. Biomass can be used as a source of energy and it most often refers to plants or plant-based materials that are not used for food or feed, and are specifically called lignocellulosic biomass. As an energy source, biomass can either be used directly via combustion to produce heat, or indirectly after converting it to various forms of biofuel. Conversion of biomass to biofuel can be achieved by different methods which are broadly classified into: *thermal*, *chemical*, and *biochemical* methods.

Biomass Sources

Eucalyptus in Brazil. Remains of the tree are reused for power generation.

Historically, humans have harnessed biomass-derived energy since the time when people began burning wood to make fire. Even today, biomass is the only source of fuel for domestic use in many developing countries. Biomass is all biologically-produced matter based in carbon, hydrogen and oxygen. The estimated biomass production in the world is 104.9 petagrams (104.9×10^{15} g – about 105 billion metric tons) of carbon per year, about half in the ocean and half on land.

Wood remains the largest biomass energy source today; examples include forest residues (such as dead trees, branches and tree stumps), yard clippings, wood chips and even municipal solid waste. Wood energy is derived by using lignocellulosic biomass (second-generation biofuels) as fuel. Harvested wood may be used directly as a fuel or collected from wood waste streams to be processed into pellet fuel or other forms of fuels. The largest source of energy from wood is pulping liquor or "black liquor, a waste product from processes of the pulp, paper and paperboard industry. In the second sense, biomass includes plant or animal matter that can be converted into fibers or other industrial chemicals, including biofuels. Industrial biomass can be grown from numerous types of plants, including miscanthus,switchgrass, hemp, corn, poplar, willow, sorghum, sugarcane, bamboo, and a variety of tree species, ranging from eucalyptus to oil palm (palm oil).

Based on the source of biomass, biofuels are classified broadly into two major categories. First-generation biofuels are derived from sources such as sugarcane and corn starch. Sugars present in this biomass are fermented to produce bioethanol, an alcohol fuel which can be used directly in a fuel cell to produce electricity or serve as an additive to gasoline. However, utilizing food-based resources for fuel production only aggravates the food shortage problem. Second-generation biofuels, on the other hand, utilize non-food-based biomass sources such as agriculture and municipal waste. These biofuels mostly consist of lignocellulosic biomass, which is not edible and is a low-value waste for many industries. Despite being the favored alternative, economical production of second-generation biofuel is not yet achieved due to technological issues. These issues arise mainly due to chemical inertness and structural rigidity of lignocellulosic biomass.

Plant energy is produced by crops specifically grown for use as fuel that offer high biomass output per hectare with low input energy. Some examples of these plants are wheat, which typically yields 7.5–8 tonnes of grain per hectare, and straw, which typically yields 3.5–5 tonnes per hectare in the UK. The grain can be used for liquid transportation fuels while the straw can be burned to produce heat or electricity. Plant biomass can also be degraded from cellulose to glucose through a series of chemical treatments, and the resulting sugar can then be used as a first-generation biofuel.

The main contributors of waste energy are municipal solid waste, manufacturing waste, and landfill gas. Energy derived from biomass is projected to be the largest non-hydroelectric renewable resource of electricity in the US between 2000 and 2020.

Biomass can be converted to other usable forms of energy like methane gas or transportation fuels like ethanol and biodiesel. Rotting garbage, and agricultural and human waste, all release methane gas, also called landfill gas or biogas. Crops such as corn and sugarcane can be fermented to produce the transportation fuel ethanol. Biodiesel, another transportation fuel, can be produced from leftover food products like vegetable oils and animal fats. Several biodiesel companies simply collect used restaurant cooking oil and convert it into biodiesel. Also, biomass-to-liquids (called BTLs) and cellulosic ethanol are still under research.

There is research involving algae or algae-derived biomass, as this non-food resource can be produced at rates five to ten times those of other types of land-based agriculture, such as corn and soy. Once harvested, it can be fermented to produce biofuels such as ethanol, butanol, and methane, as well as biodiesel and hydrogen. Efforts are being made to identify which species of algae are most suitable for energy production. Genetic engineering approaches could also be utilized to improve microalgae as a source of biofuel.

The biomass used for electricity generation varies by region. Forest by-products, such as wood residues, are common in the US. Agricultural waste is common in Mauritius (sugar cane residue) and Southeast Asia (rice husks). Animal husbandry residues, such as poultry litter, are common in the UK.

As of 2015, a new bioenergy sewage treatment process aimed at developing countries is under trial; the Omni Processor is a self-sustaining process which uses sewerage solids as fuel in a process to convert waste water into drinking water, with surplus electrical energy being generated for export.

Comparison of Total Plant Biomass Yields (dry basis)

World Resources

If the total annual primary production of biomass is just over 100 billion (1.0E+11) tonnes of Carbon /yr, and the energy reserve per tonne of biomass is between about 1.5×10^3 and 3×10^3 kilowatt hours (5×10^6 and 10×10^6 BTU), or 24.8 TW average, then biomass could in principle provide 1.4 times the approximate annual 150×10^3 terawatt-hours required for current world energy consumption. For reference, the total solar power on Earth is 174 PW. The biomass equivalent to solar energy ratio is 143 ppm (parts per million), given current living system coverage on Earth. The best currently attainable solar cell efficiency is 20–40%. Additionally, Earth's internal radioactive energy production, largely the driver for volcanic activity, continental drift, etc., is in the same range of power, 20 TW. At around 50% carbon mass content in biomass, annual production, this corresponds to about 6% atmospheric carbon content in the form of CO_2 (for the current 400 ppm).

(1×10^{11} tonnes biomass annually produced approximately 25 TW·h)

Annual world biomass energy equivalent =16.7–33.4 TW·h.

Annual world energy consumption =17.7 TW·h. On average, biomass production is 1.4 times larger than world energy consumption.

Common Commodity Food Crops

- Agave: 1–21 tons/acre
- Alfalfa: 4–6 tons/acre
- Barley: grains – 1.6–2.8 tons/acre, straw – 0.9–2.5 tons/acre, total – 2.5–5.3 tons/acre
- Corn: grains – 3.2–4.9 tons/acre, stalks and stovers – 2.3–3.4 tons/acre, total – 5.5–8.3 tons/acre
- Jerusalem artichokes: tubers 1–8 tons/acre, tops 2–13 tons/acre, total 9–13 tons/acre
- Oats: grains – 1.4–5.4 tons/acre, straw – 1.9–3.2 tons/acre, total – 3.3–8.6 tons/acre
- Rye: grains – 2.1–2.4 tons/acre, straw – 2.4–3.4 tons/acre, total – 4.5–5.8 tons/acre
- Wheat: grains – 1.2–4.1 tons/acre, straw – 1.6–3.8 tons/acre, total – 2.8–7.9 tons/acre

Woody Crops

- Oil palm: fronds 11 ton/acre, whole fruit bunches 1 ton/acre, trunks 30 ton/acre

Not Yet In Commercial Planting

- Giant miscanthus: 5–15 tons/acre
- Sunn hemp: 4.5 tons/acre
- Switchgrass: 4–6 tons/acre

Genetically Modified Varieties

- Energy Sorghum

Biomass Conversion

Thermal Conversions

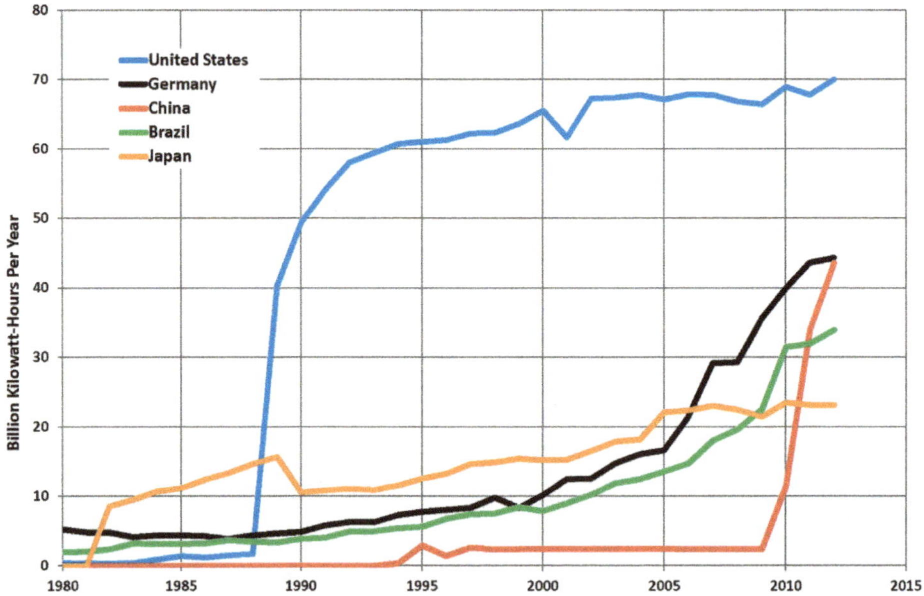

Trends in the top five countries generating electricity from biomass.

Biomass briquettes are an example fuel for production of dendrothermal energy.

Thermal conversion processes use heat as the dominant mechanism to convert biomass into another chemical form. Also known as thermal oil heating, it is a type of indirect heating in which a liquid phase heat transfer medium is heated and circulated to one or more heat energy users within a closed loop system.The basic alternatives of combustion (torrefaction, pyrolysis, and gasification) are separated principally by the extent to which the chemical reactions involved are allowed to proceed (mainly controlled by the availability of oxygen and conversion temperature).

Energy created by burning biomass is particularly suited for countries where the fuel wood grows more rapidly, e.g. tropical countries. There are a number of other less common, more experimental or proprietary thermal processes that may offer benefits such as hydrothermal upgrading (HTU) and hydroprocessing. Some have been developed for use on high moisture content biomass, including aqueous slurries, and allow them to be converted into more convenient forms. Some of the applications of thermal conversion are combined heat and power (CHP) and co-firing. In a typical dedicated biomass power plant, efficiencies range from 20–27% (higher heating value basis). Biomass cofiring with coal, by contrast, typically occurs at efficiencies near those of the coal combustor (30–40%, higher heating value basis).

Chemical Conversion

A range of chemical processes may be used to convert biomass into other forms, such as to produce a fuel that is more conveniently used, transported or stored, or to exploit some property of the process itself. Many of these processes are based in large part on similar coal-based processes, such as Fischer-Tropsch synthesis, methanol production, olefins (ethylene and propylene), and similar chemical or fuel feedstocks. In most cases, the first step involves gasification, which step generally is the most expensive and involves the greatest technical risk. Biomass is more difficult to feed into a pressure vessel than coal or any liquid. Therefore, biomass gasification is frequently done at atmospheric pressure and causes combustion of biomass to produce a combustible gas consisting of carbon monoxide, hydrogen, and traces of methane. This gas mixture, called a producer gas, can provide fuel for various vital processes, such as internal combustion engines, as well as substitute for furnace oil in direct heat applications. Because any biomass material can undergo gasification, this process is far more attractive than ethanol or biomass production, where only particular biomass materials can be used to produce a fuel. In addition, biomass gasification is a desirable process due to the ease at which it can convert solid waste (such as wastes available on a farm) into producer gas, which is a very usable fuel.

Conversion of biomass to biofuel can also be achieved via selective conversion of individual components of biomass. For example, cellulose can be converted to intermediate platform chemical such a sorbitol, glucose, hydroxymethylfurfural etc. These chemical are then further reacted to produce hydrogen or hydrocarbon fuels.

Biomass also has the potential to be converted to multiple commodity chemicals. Halomethanes have successfully been by produced using a combination of A. fermentans and engineered S. cerevisiae. This method converts NaX salts and unprocessed biomass such as switchgrass, sugarcane, corn stover, or poplar into halomethanes. S-adenosylmethionine which is naturally occurring in S. cerevisiae allows a methyl group to be transferred. Production levels of 150 mg L-1H-1 iodomethane were achieved. At these levels roughly 173000L of capacity would need to be operated just to replace the United States' need for iodomethane. However, an advantage of this method is that it uses NaI rather than I2; NaI is significantly less hazardous than I2. This method may be applied to produce ethylene in the future.

Other chemical processes such as converting straight and waste vegetable oils into biodiesel is transesterification.

Biochemical Conversion

As biomass is a natural material, many highly efficient biochemical processes have developed in nature to break down the molecules of which biomass is composed, and many of these biochemical conversion processes can be harnessed.

Biochemical conversion makes use of the enzymes of bacteria and other microorganisms to break down biomass into gaseous or liquid fuels, such a biogas or bioethanol. In most cases, microorganisms are used to perform the conversion process: anaerobic digestion, fermentation, and composting.

Electrochemical Conversion

In addition to combustion, biomass/biofuels can be directly converted to electrical energy via electrochemical (electrocatalytic) oxidation of the material. This can be performed directly in a direct carbon fuel cell, direct liquid fuel cells such as direct ethanol fuel cell, a direct methanol fuel cell, a direct formic acid fuel cell, a L-ascorbic Acid Fuel Cell (vitamin C fuel cell), and a microbial fuel cell. The fuel can also be consumed indirectly via a fuel cell system containing a reformer which converts the biomass into a mixture of CO and H2 before it is consumed in the fuel cell.

In the United States

The biomass power generating industry in the United States consists of approximately 11,000 MW of summer operating capacity actively supplying power to the grid, and produces about 1.4 percent of the U.S. electricity supply.

Public Service of New Hampshire (later merged with other companies into Eversource) in 2006 replaced a 50 MW coal boiler with a new 50 MW biomass boiler at its Schiller Station facility in Portsmouth, NH. The boiler's biomass fuel is from sources in NH, Massachusetts and Maine.

Currently, the New Hope Power Partnership is the largest biomass power plant in the U.S. The 140 MW facility uses sugarcane fiber (bagasse) and recycled urban wood as fuel to generate enough power for its large milling and refining operations as well as to supply electricity for nearly 60,000 homes.

Second-Generation Biofuels

Second-generation biofuels were not (in 2010) produced commercially, but a considerable number of research activities were taking place mainly in North America, Europe and also in some emerging countries. These tend to use feedstock produced by rapidly reproducing enzymes or bacteria from various sources including excrement grown in cell cultures or hydroponics. There is huge potential for second generation biofuels but non-edible feedstock resources are highly under-utilized.

Environmental Impact

Using biomass as a fuel produces air pollution in the form of carbon monoxide, carbon dioxide, NOx (nitrogen oxides), VOCs (volatile organic compounds), particulates and other pollutants at levels above those from traditional fuel sources such as coal or natural gas in some cases (such as with indoor heating and cooking). Utilization of wood biomass as a fuel can also produce fewer particulate and other pollutants than open burning as seen in wildfires or direct heat applications. Black carbon – a pollutant created by combustion of fossil fuels, biofuels, and biomass – is possibly the second largest contributor to global warming.[:56–57] In 2009 a Swedish study of the giant brown haze that periodically covers large areas in South Asia determined that it had been principally produced by open burning of biomass, and to a lesser extent by fossil-fuel burning. Researchers measured a significant concentration of ^{14}C (Carbon-14), which is associated with recent plant life rather than with fossil fuels.

Biomass power plant size is often driven by biomass availability in close proximity as transport costs of the (bulky) fuel play a key factor in the plant's economics. It has to be noted, however, that rail and especially shipping on waterways can reduce transport costs significantly, which has led to a global biomass market. To make small plants of 1 MW_{el} economically profitable those power plants need to be equipped with technology that is able to convert biomass to useful electricity with high efficiency such as ORC technology, a cycle similar to the water steam power process just with an organic working medium. Such small power plants can be found in Europe.

On combustion, the carbon from biomass is released into the atmosphere as carbon dioxide (CO_2). The amount of carbon stored in dry wood is approximately 50% by weight. However, according to the Food and Agriculture Organization of the United Nations, plant matter used as a fuel can be replaced by planting for new growth. When the biomass is from forests, the time to recapture the carbon stored is generally longer, and the carbon storage capacity of the forest may be reduced overall if destructive forestry techniques are employed.

Industry professionals claim that a range of issues can affect a plant's ability to comply with emissions standards. Some of these challenges, unique to biomass plants, include inconsistent fuel supplies and age. The type and amount of the fuel supply are completely reliant factors; the fuel can be in the form of building debris or agricultural waste (such as removal of invasive species or orchard trimmings). Furthermore, many of the biomass plants are old, use outdated technology and were not built to comply with today's stringent standards. In fact, many are based on technologies developed during the term of U.S. President Jimmy Carter, who created the United States Department of Energy in 1977.

The U.S. Energy Information Administration projected that by 2017, biomass is expected to be about twice as expensive as natural gas, slightly more expensive than nuclear power, and much less expensive than solar panels. In another EIA study released, concerning the government's plan to implement a 25% renewable energy standard by 2025, the agency assumed that 598 million tons of biomass would be available, accounting for 12% of the renewable energy in the plan.

The adoption of biomass-based energy plants has been a slow but steady process. Between the years of 2002 and 2012 the production of these plants has increased 14%. In the United States, alternative electricity-production sources on the whole generate about 13% of power; of this fraction, biomass contributes approximately 11% of the alternative production. According to a study conducted in early 2012, of the 107 operating biomass plants in the United States, 85 have been cited by federal or state regulators for the violation of clean air or water standards laws over the past 5 years. This data also includes minor infractions.

Despite harvesting, biomass crops may sequester carbon. For example, soil organic carbon has been observed to be greater in switchgrass stands than in cultivated cropland soil, especially at depths below 12 inches. The grass sequesters the carbon in its increased root biomass. Typically, perennial crops sequester much more carbon than annual crops due to much greater non-harvested living biomass, both living and dead, built up over years, and much less soil disruption in cultivation.

The proposal that biomass is carbon-neutral put forward in the early 1990s has been superseded by more recent science that recognizes that mature, intact forests sequester carbon more effectively than cut-over areas. When a tree's carbon is released into the atmosphere in a single pulse, it contributes to climate change much more than woodland timber rotting slowly over decades. Current studies indicate that even after 50 years the forest has not recovered to its initial carbon storage and the optimal strategy is likely to be protection of the standing forest".

The pros and cons of biomass usage regarding carbon emissions may be quantified with the ILUC factor. There is controversy surrounding the usage of the ILUC factor.

Forest-based biomass has recently come under fire from a number of environmental organizations, including Greenpeace and the Natural Resources Defense Council, for the harmful impacts it can have on forests and the climate. Greenpeace recently released a report entitled Fuelling a BioMess which outlines their concerns around forest-based biomass. Because any part of the tree can be burned, the harvesting of trees for energy production encourages whole-tree harvesting, which removes more nutrients and soil cover than regular harvesting, and can be harmful to the long-term health of the forest. In some jurisdictions, forest biomass removal is increasingly involving elements essential to functioning forest ecosystems, including standing trees, naturally disturbed forests and remains of traditional logging operations that were previously left in the forest. Environmental groups also cite recent scientific research which has found that it can take many decades for the carbon released by burning biomass to be recaptured by regrowing trees, and even longer in low productivity areas; furthermore, logging operations may disturb forest soils and cause them to release stored carbon. In light of the pressing need to reduce greenhouse gas emissions in the short term in order to mitigate the effects of climate change, a number of environmental groups are opposing the large-scale use of forest biomass in energy production.

Supply Chain Issues

With the seasonality of biomass supply and a great variability in sources, supply chains play a key role in cost-effective delivery of bioenergy. There are several potential challenges unique to bioenergy supply chains:

Technical issues

- Inefficiencies of the conversion processes
- Storage methods for seasonal availability
- Complex multi-component constituents incompatible with maximizing efficiency of single purpose use
- High water content of many biomass feedstock
- Conflicting decisions (technologies, locations, and routes)
- Complex location analysis (source points, inventory facilities, and production plants)

Logistic issues

- Seasonal availability leading to storage challenges and/or seasonally idle facilities
- Low bulk-density and/or high water content making transportation of biomass less economical
- Finite productivity per area and/or time incompatible with conventional approach to economy of scale focusing on maximizing facility size

Financial issues

- The limits for the traditional approach to economy of scale which focuses on maximizing single facility size
- Unavailability and complexity of life cycle costing data
- Lack of required transport infrastructure
- Limited flexibility or inflexibility to energy demand
- Risks associated with new technologies (insurability, performance, rate of return)
- Extended market volatilities (conflicts with alternative markets for biomass)
- Difficult or impossible to use financial hedging methods to control cost

Social issues

- Lack of participatory decision making
- Lack of public/community awareness
- Local supply chain impacts vs. global benefits
- Health and safety risks

- Extra pressure on transport sector
- Decreasing the esthetics of rural areas

Policy and regulatory issues

- Impact of fossil fuel tax on biomass transport
- Lack of incentives to create competition among bioenergy producers
- Focus on technology options and less attention to selection of biomass materials
- Lack of support for sustainable supply chain solutions

Institutional and organizational issues

- Varied ownership arrangements and priorities among supply chain parties
- Lack of supply chain standards
- Impact of organizational norms and rules on decision making and supply chain coordination
- Immaturity of change management practices in biomass supply chains

BIOFUEL

A **biofuel** is a fuel that is produced through contemporary biological processes, such as agriculture and anaerobic digestion, rather than a fuel produced by geological processes such as those involved in the formation of fossil fuels, such as coal and petroleum, from prehistoric biological matter. Biofuels can be derived directly from plants, or indirectly from agricultural, commercial, domestic, and/or industrial wastes. Renewable biofuels generally involve contemporary carbon fixation, such as those that occur in plants or microalgae through the process of photosynthesis. Other renewable biofuels are made through the use or conversion of biomass (referring to recently living organisms, most often referring to plants or plant-derived materials). This biomass can be converted to convenient energy-containing substances in three different ways: thermal conversion, chemical conversion, and biochemical conversion. This biomass conversion can result in fuel in solid, liquid, or gas form. This new biomass can also be used directly for biofuels.

A bus fueled by biodiesel.

Information on pump regarding ethanol fuel blend up to 10%, California.

Bioethanol is an alcohol made by fermentation, mostly from carbohydrates produced in sugar or starch crops such as corn, sugarcane, or sweet sorghum. Cellulosic biomass, derived from non-food sources, such as trees and grasses, is also being developed as a feedstock for ethanol production. Ethanol can be used as a fuel for vehicles in its pure form, but it is usually used as a gasoline additive to increase octane and improve vehicle emissions. Bioethanol is widely used in the USA and in Brazil. Current plant design does not provide for converting the lignin portion of plant raw materials to fuel components by fermentation.

Biodiesel can be used as a fuel for vehicles in its pure form, but it is usually used as a diesel additive to reduce levels of particulates, carbon monoxide, and hydrocarbons from diesel-powered vehicles. Biodiesel is produced from oils or fats using transesterification and is the most common biofuel in Europe.

In 2010, worldwide biofuel production reached 105 billion liters (28 billion gallons US), up 17% from 2009, and biofuels provided 2.7% of the world's fuels for road transport. Global ethanol fuel production reached 86 billion liters (23 billion gallons US) in 2010, with the United States and Brazil as the world's top producers, accounting together for about 90% of global production. The world's largest biodiesel producer is the European Union, accounting for 53% of all biodiesel production in 2010. As of 2011, mandates for blending biofuels exist in 31 countries at the national level and in 29 states or provinces. The International Energy Agency has a goal for biofuels to meet more than a quarter of world demand for transportation fuels by 2050 to reduce dependence on petroleum and coal. The production of biofuels also led into a flourishing automotive industry, where by 2010, 79% of all cars produced in Brazil were made with a hybrid fuel system of bioethanol and gasoline.

There are various social, economic, environmental and technical issues relating to biofuels production and use, which have been debated in the popular media and scientific journals. These include: the effect of moderating oil prices, the "food vs fuel" debate, poverty reduction potential, carbon emissions levels, sustainable biofuel production, deforestation and soil erosion, loss of biodiversity, impact on water resources, rural social exclusion and injustice, shantytown migration, rural unskilled unemployment, and nitrous oxide (NO2) emissions.

Liquid Fuels for Transportation

Most transportation fuels are liquids, because vehicles usually require high energy density. This occurs naturally in liquids and solids. High energy density can also be provided by an internal combustion engine. These engines require clean-burning fuels. The fuels that are easiest to burn cleanly are typically liquids and gases. Thus, liquids meet the requirements of being both energy-dense and clean-burning. In addition, liquids (and gases) can be pumped, which means handling is easily mechanized, and thus less laborious.

First-Generation Biofuels

"First-generation" or conventional biofuels are made from sugar, starch, or vegetable oil.

Ethanol

Neat ethanol on the left (A), gasoline on the right (G) at a filling station in Brazil.

Biologically produced alcohols, most commonly ethanol, and less commonly propanol and butanol, are produced by the action of microorganisms and enzymes through the fermentation of sugars or starches (easiest), or cellulose (which is more difficult). Biobutanol (also called biogasoline) is often claimed to provide a direct replacement for gasoline, because it can be used directly in a gasoline engine.

U.S. President George W. Bush looks at sugar cane, a source of biofuel, with Brazilian President Luiz Inácio Lula da Silva during a tour on biofuel technology at Petrobras in São Paulo, Brazil, 9 March 2007.

Ethanol fuel is the most common biofuel worldwide, particularly in Brazil. Alcohol fuels are produced by fermentation of sugars derived from wheat, corn, sugar beets, sugar cane, molasses and any sugar or starch from which alcoholic beverages such as whiskey, can be made (such as potato and fruit waste, etc.). The ethanol production methods used are enzyme digestion (to release sugars from stored starches), fermentation of the sugars, distillation and drying. The distillation process requires significant energy input for heat (sometimes unsustainable natural gas fossil fuel, but cellulosic biomass such as bagasse, the waste left after sugar cane is pressed to extract its juice, is the most common fuel in Brazil, while pellets, wood chips and also waste heat are more common in Europe) Waste steam fuels ethanol factory - where waste heat from the factories also is used in the district heating grid.

Ethanol can be used in petrol engines as a replacement for gasoline; it can be mixed with gasoline to any percentage. Most existing car petrol engines can run on blends of up to 15% bioethanol with petroleum/gasoline. Ethanol has a smaller energy density than that of gasoline; this means it takes more fuel (volume and mass) to produce the same amount of work. An advantage of ethanol (CH_3CH_2OH) is that it has a higher octane rating than ethanol-free gasoline available at roadside gas stations, which allows an increase of an engine's compression ratio for increased thermal efficiency. In high-altitude (thin air) locations, some states mandate a mix of gasoline and ethanol as a winter oxidizer to reduce atmospheric pollution emissions.

Ethanol is also used to fuel bioethanol fireplaces. As they do not require a chimney and are "flueless", bioethanol fires are extremely useful for newly built homes and apartments without a flue. The downsides to these fireplaces is that their heat output is slightly less than electric heat or gas fires, and precautions must be taken to avoid carbon monoxide poisoning.

Corn-to-ethanol and other food stocks has led to the development of cellulosic ethanol. According to a joint research agenda conducted through the US Department of Energy, the fossil energy ratios (FER) for cellulosic ethanol, corn ethanol, and gasoline are 10.3, 1.36, and 0.81, respectively.

Ethanol has roughly one-third lower energy content per unit of volume compared to gasoline. This is partly counteracted by the better efficiency when using ethanol (in a long-term test of more than 2.1 million km, the BEST project found FFV vehicles to be 1-26 % more energy efficient than petrol cars, but the volumetric consumption increases by approximately 30%, so more fuel stops are required.

With current subsidies, ethanol fuel is slightly cheaper per distance traveled in the United States.

Biodiesel

Biodiesel is the most common biofuel in Europe. It is produced from oils or fats using transesterification and is a liquid similar in composition to fossil/mineral diesel. Chemically, it consists mostly of fatty acid methyl (or ethyl) esters (FAMEs). Feedstocks for biodiesel include animal fats, vegetable oils, soy, rapeseed, jatropha, mahua, mustard, flax, sunflower, palm oil, hemp, field pennycress, *Pongamia pinnata* and algae. Pure biodiesel (B100) currently reduces emissions with up to 60% compared to diesel Second generation B100.

Biodiesel can be used in any diesel engine when mixed with mineral diesel. In some countries, manufacturers cover their diesel engines under warranty for B100 use, although Volkswagen of Germany, for example, asks drivers to check by telephone with the VW environmental services department before switching to B100. B100 may become more viscous at lower temperatures, depending on the feedstock used. In most cases, biodiesel is compatible with diesel engines from 1994 onwards, which use 'Viton' (by DuPont) synthetic rubber in their mechanical fuel injection systems. Note however, that no vehicles are certified for using neat biodiesel before 2014, as there was no emission control protocol available for biodiesel before this date.

Electronically controlled 'common rail' and 'unit injector' type systems from the late 1990s onwards may only use biodiesel blended with conventional diesel fuel. These engines have finely metered and atomized multiple-stage injection systems that are very sensitive to the viscosity of the fuel. Many current-generation diesel engines are made so that they can run on B100 without altering the engine itself, although this depends on the fuel rail design. Since biodiesel is an effective solvent and cleans residues deposited by mineral diesel, engine filters may need to be replaced more often, as the biofuel dissolves old deposits in the fuel tank and pipes. It also effectively cleans the engine combustion chamber of carbon deposits, helping to maintain efficiency. In many European countries, a 5% biodiesel blend is widely used and is available at thousands of gas stations. Biodiesel is also an oxygenated fuel, meaning it contains a reduced amount of carbon and higher hydrogen and oxygen content than fossil diesel. This improves the combustion of biodiesel and reduces the particulate emissions from unburnt carbon. However, using neat biodiesel may increase NOx-emissions

Biodiesel is also safe to handle and transport because it is non-toxic and biodegradable, and has a high flash point of about 300 °F (148 °C) compared to petroleum diesel fuel, which has a flash point of 125 °F (52 °C).

In the USA, more than 80% of commercial trucks and city buses run on diesel. The emerging US biodiesel market is estimated to have grown 200% from 2004 to 2005. By the end of 2006 biodiesel production was estimated to increase fourfold [from 2004] to more than 1 billion US gallons (3,800,000 m^3).

In France, biodiesel is incorporated at a rate of 8% in the fuel used by all French diesel vehicles. Avril Group produces under the brand Diester, a fifth of 11 million tons of biodiesel consumed annually by the European Union. It is the leading European producer of biodiesel.

Other Bioalcohols

Methanol is currently produced from natural gas, a non-renewable fossil fuel. In the future it is hoped to be produced from biomass as biomethanol. This is technically feasible, but the production is currently being postponed for concerns of Jacob S. Gibbs and Brinsley Coleberd that the economic viability is still pending. The methanol economy is an alternative to the hydrogen economy, compared to today's hydrogen production from natural gas.

Butanol (C_4H_9OH) is formed by ABE fermentation (acetone, butanol, ethanol) and experimental modifications of the process show potentially high net energy gains with butanol as the only liquid product. Butanol will produce more energy and allegedly can be burned "straight" in existing gasoline engines (without modification to the engine or car), and is less corrosive and less water-soluble than ethanol, and could be distributed via existing infrastructures. DuPont and BP are working together to help develop butanol. *E. coli* strains have also been successfully engineered to produce butanol by modifying their amino acid metabolism.

Green Diesel

Green diesel is produced through hydrocracking biological oil feedstocks, such as vegetable oils and animal fats. Hydrocracking is a refinery method that uses elevated temperatures and pressure in the presence of a catalyst to break down larger molecules, such as those found in vegetable oils, into shorter hydrocarbon chains used in diesel engines. It may also be called renewable diesel, hydrotreated vegetable oil or hydrogen-derived renewable diesel. Green diesel has the same chemical properties as petroleum-based diesel. It does not require new engines, pipelines or infrastructure to distribute and use, but has not been produced at a cost that is competitive with petroleum. Gasoline versions are also being developed. Green diesel is being developed in Louisiana and Singapore by ConocoPhillips, Neste Oil, Valero, Dynamic Fuels, and Honeywell UOP as well as Preem in Gothenburg, Sweden, creating what is known as Evolution Diesel.

Biofuel Gasoline

In 2013 UK researchers developed a genetically modified strain of Escherichia coli (E.Coli), which could transform glucose into biofuel gasoline that does not need to be blended. Later in 2013 UCLA researchers engineered a new metabolic pathway to bypass glycolysis and increase the rate of conversion of sugars into biofuel, while KAIST researchers developed a strain capable of producing short-chain alkanes, free fatty acids, fatty esters and fatty alcohols through the fatty acyl (acyl carrier protein (ACP)) to fatty acid to fatty acyl-CoA pathway *in vivo*. It is believed that in the future it will be possible to "tweak" the genes to make gasoline from straw or animal manure.

Vegetable Oil

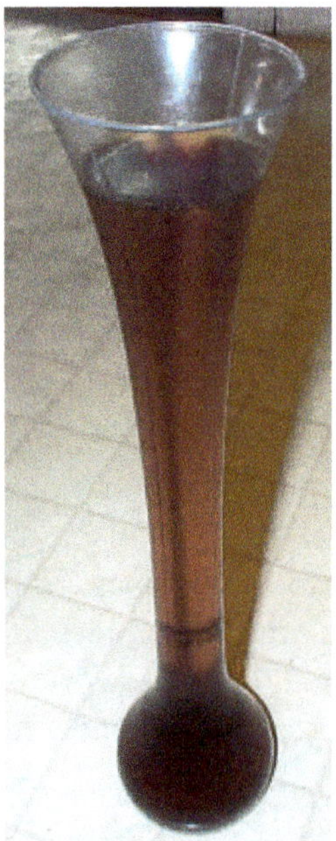

Filtered waste vegetable oil.

Straight unmodified edible vegetable oil is generally not used as fuel, but lower-quality oil can and has been used for this purpose. Used vegetable oil is increasingly being processed into biodiesel, or (more rarely) cleaned of water and particulates and used as a fuel.

As with 100% biodiesel (B100), to ensure the fuel injectors atomize the vegetable oil in the correct pattern for efficient combustion, vegetable oil fuel must be heated to reduce its viscosity to that of diesel, either by electric coils or heat exchangers. This is easier in warm or temperate climates. MAN B&W Diesel, Wärtsilä, and Deutz AG, as well as a number of smaller companies, such as Elsbett, offer engines that are compatible with straight vegetable oil, without the need for after-market modifications.

Walmart's truck fleet logs millions of miles each year, and the company planned to double the fleet's efficiency between 2005 and 2015. This truck is one of 15 based at Walmart's Buckeye, Arizona distribution center that was converted to run on a biofuel made from reclaimed cooking grease produced during food preparation at Walmart stores.

Vegetable oil can also be used in many older diesel engines that do not use common rail or unit injection electronic diesel injection systems. Due to the design of the combustion chambers in indirect injection engines, these are the best engines for use with vegetable oil. This system allows the relatively larger oil molecules more time to burn. Some older engines, especially Mercedes, are driven experimentally by enthusiasts without any conversion, a handful of drivers have experienced limited success with earlier pre-"Pumpe Duse" VW TDI engines and other similar engines with direct injection. Several companies, such as Elsbett or Wolf, have developed professional conversion kits and successfully installed hundreds of them over the last decades.

Oils and fats can be hydrogenated to give a diesel substitute. The resulting product is a straight-chain hydrocarbon with a high cetane number, low in aromatics and sulfur and does not contain oxygen. Hydrogenated oils can be blended with diesel in all proportions. They have several advantages over biodiesel, including good performance at low temperatures, no storage stability problems and no susceptibility to microbial attack.

Bioethers

Bioethers (also referred to as fuel ethers or oxygenated fuels) are cost-effective compounds that act as octane rating enhancers."Bioethers are produced by the reaction of reactive iso-olefins, such as iso-butylene, with bioethanol.Bioethers are created by wheat or sugar beet. They also enhance engine performance, whilst significantly reducing engine wear and toxic exhaust emissions. Though bioethers are likely to replace petroethers in the UK, it is highly unlikely they will become a fuel in and of itself due to the low energy density. Greatly reducing the amount of ground-level ozone emissions, they contribute to air quality.

When it comes to transportation fuel there are six ether additives: dimethyl ether (DME), diethyl ether (DEE), methyl teritiary-butyl ether (MTBE), ethyl *ter*-butyl ether (ETBE), t*er*-amyl methyl ether (TAME), and *ter*-amyl ethyl ether (TAEE)

The European Fuel Oxygenates Association (EFOA) credits methyl Ttertiary-butyl ether (MTBE) and ethyl ter-butyl ether (ETBE) as the most commonly used ethers in fuel to replace lead. Ethers were introduced in Europe in the 1970s to replace the highly toxic compound. Although Europeans still use bio-ether additives, the US no longer has an oxygenate requirement therefore bio-ethers are no longer used as the main fuel additive.

Biogas

Pipes carrying biogas.

Biogas is methane produced by the process of anaerobic digestion of organic material by anaerobes. It can be produced either from biodegradable waste materials or by the use of energy crops fed into anaerobic digesters to supplement gas yields. The solid byproduct, digestate, can be used as a biofuel or a fertilizer.

Biogas can be recovered from mechanical biological treatment waste processing systems. Landfill gas, a less clean form of biogas, is produced in landfills through naturally occurring anaerobic digestion. If it escapes into the atmosphere, it is a potential greenhouse gas.

Farmers can produce biogas from manure from their cattle by using anaerobic digesters.

Syngas

Syngas, a mixture of carbon monoxide, hydrogen and other hydrocarbons, is produced by partial combustion of biomass, that is, combustion with an amount of oxygen that is not sufficient to convert the biomass completely to carbon dioxide and water. Before partial combustion, the biomass is dried, and sometimes pyrolysed. The resulting gas mixture, syngas, is more efficient than direct combustion of the original biofuel; more of the energy contained in the fuel is extracted.

Syngas may be burned directly in internal combustion engines, turbines or high-temperature fuel cells. The wood gas generator, a wood-fueled gasification reactor, can be connected to an internal combustion engine.

Syngas can be used to produce methanol, DME and hydrogen, or converted via the Fischer-Tropsch process to produce a diesel substitute, or a mixture of alcohols that can be blended into gasoline. Gasification normally relies on temperatures greater than 700 °C.

Lower-temperature gasification is desirable when co-producing biochar, but results in syngas polluted with tar.

Solid Biofuels

Examples include wood, sawdust, grass trimmings, domestic refuse, charcoal, agricultural waste, nonfood energy crops, and dried manure.

When raw biomass is already in a suitable form (such as firewood), it can burn directly in a stove or furnace to provide heat or raise steam. When raw biomass is in an inconvenient form (such as sawdust, wood chips, grass, urban waste wood, agricultural residues), the typical process is to densify the biomass. This process includes grinding the raw biomass to an appropriate particulate size (known as hogfuel), which, depending on the densification type, can be from 1 to 3 cm (0.4 to 1.2 in), which is then concentrated into a fuel product. The current processes produce wood pellets, cubes, or pucks. The pellet process is most common in Europe, and is typically a pure wood product. The other types of densification are larger in size compared to a pellet, and are compatible with a broad range of input feedstocks. The resulting densified fuel is easier to transport and feed into thermal generation systems, such as boilers.

Industry has used sawdust, bark and chips for fuel for decades, primary in the pulp and paper industry, and also bagasse (spent sugar cane) fueled boilers in the sugar cane industry. Boilers in the range of 500,000 lb/hr of steam, and larger, are in routine operation, using grate, spreader stoker, suspension burning and fluid bed combustion. Utilities generate power, typically in the range of 5 to 50 MW, using locally available fuel. Other industries have also installed wood waste fueled boilers and dryers in areas with low cost fuel.

One of the advantages of biomass fuel is that it is often a byproduct, residue or waste-product of other processes, such as farming, animal husbandry and forestry. In theory, this means fuel and food production do not compete for resources, although this is not always the case.

A problem with the combustion of raw biomass is that it emits considerable amounts of pollutants, such as particulates and polycyclic aromatic hydrocarbons. Even modern pellet boilers generate much more pollutants than oil or natural gas boilers. Pellets made from agricultural residues are usually worse than wood pellets, producing much larger emissions of dioxins and chlorophenols.

In spite of the above noted study, numerous studies have shown biomass fuels have significantly less impact on the environment than fossil based fuels. Of note is the US Department of Energy Laboratory, operated by Midwest Research Institute Biomass Power and Conventional Fossil Systems with and without CO2 Sequestration – Comparing the Energy Balance, Greenhouse Gas Emissions and Economics Study. Power generation emits significant amounts of greenhouse gases (GHGs), mainly carbon dioxide (CO_2). Sequestering CO_2 from the power plant flue gas can significantly reduce the GHGs from the power plant itself, but this is not the total picture. CO_2 capture and sequestration consumes additional energy, thus lowering the plant's fuel-to-electricity efficiency. To compensate for this, more fossil fuel must be procured and consumed to make up for lost capacity.

Taking this into consideration, the global warming potential (GWP), which is a combination of CO_2, methane (CH_4), and nitrous oxide (N_2O) emissions, and energy balance of the system need to be examined using a life cycle assessment. This takes into account the upstream processes which remain constant after CO_2sequestration, as well as the steps required for additional power generation. Firing biomass instead of coal led to a 148% reduction in GWP.

A derivative of solid biofuel is biochar, which is produced by biomass pyrolysis. Biochar made from agricultural waste can substitute for wood charcoal. As wood stock becomes scarce, this alternative is gaining ground. In eastern Democratic Republic of Congo, for example, biomass briquettes are being marketed as an alternative to charcoal to protect Virunga National Park from deforestation associated with charcoal production.

Second-Generation (advanced) Biofuels

Second generation biofuels, also known as advanced biofuels, are fuels that can be manufactured from various types of biomass. Biomass is a wide-ranging term meaning any source of organic carbon that is renewed rapidly as part of the carbon cycle. Biomass is derived from plant materials but can also include animal materials.

First generation biofuels are made from the sugars and vegetable oils found in arable crops, which can be easily extracted using conventional technology. In comparison, second generation biofuels are made from lignocellulosic biomass or woody crops, agricultural residues or waste, which makes it harder to extract the required fuel. A series of physical and chemical treatments might be required to convert lignocellulosic biomass to liquid fuels suitable for transportation.

Sustainable Biofuels

Biofuels in the form of liquid fuels derived from plant materials are entering the market, driven mainly by the perception that they reduce climate gas emissions, and also by factors such as oil price spikes and the need for increased energy security. However, many of the biofuels that are currently being supplied have been criticised for their adverse impacts on the natural environment, food security, and land use. In 2008, the Nobel-prize winning chemist Paul J. Crutzen published findings that the release of nitrous oxide (N_2O) emissions in the production of biofuels means that overall they contribute more to global warming than the fossil fuels they replace.

The challenge is to support biofuel development, including the development of new cellulosic technologies, with responsible policies and economic instruments to help ensure that biofuel commercialization is sustainable. Responsible commercialization of biofuels represents an opportunity to enhance sustainable economic prospects in Africa, Latin America and Asia.

According to the Rocky Mountain Institute, sound biofuel production practices would not hamper food and fibre production, nor cause water or environmental problems, and would enhance soil fertility. The selection of land on which to grow the feedstocks is a critical component of the ability of biofuels to deliver sustainable solutions. A key consideration is the minimisation of biofuel competition for prime cropland.

Biofuels by Region

Bio Diesel Powered Fast Attack Craft Of Indian Navy patrolling during IFR 2016.The green bands on the vessels are indicative of the fact that the vessels are powered by bio-diesel.

There are international organizations such as IEA Bioenergy, established in 1978 by the OECD International Energy Agency (IEA), with the aim of improving cooperation and information exchange between countries that have national programs in bioenergy research, development and deployment. The UN International Biofuels Forum is formed by Brazil, China, India, Pakistan, South Africa, the United States and the European Commission. The world leaders in biofuel development and use are Brazil, the United States, France, Sweden and Germany. Russia also has 22% of world's forest, and is a big biomass (solid biofuels) supplier. In 2010, Russian pulp and paper maker, Vyborgskaya Cellulose, said they would be producing pellets that can be used in heat and electricity generation from its plant in Vyborg by the end of the year. The plant will eventually produce about 900,000 tons of pellets per year, making it the largest in the world once operational.

Biofuels currently make up 3.1% of the total road transport fuel in the UK or 1,440 million litres. By 2020, 10% of the energy used in UK road and rail transport must come from renewable sources – this is the equivalent of replacing 4.3 million tonnes of fossil oil each year. Conventional biofuels are likely to produce between 3.7 and 6.6% of the energy needed in road and rail transport, while advanced biofuels could meet up to 4.3% of the UK's renewable transport fuel target by 2020.

Air Pollution

Biofuels are different from fossil fuels in regard to greenhouse gases but are similar to fossil fuels in that biofuels contribute to air pollution. Burning produces airborne carbon particulates, carbon monoxide and nitrous oxides. The WHO estimates 3.7 million premature deaths worldwide in 2012 due to air pollution.Brazil burns significant amounts of ethanol biofuel. Gas chromatograph studies were performed of ambient air in São Paulo, Brazil, and compared to Osaka, Japan, which does not burn ethanol fuel. Atmospheric Formaldehyde was 160% higher in Brazil, and Acetaldehyde was 260% higher.

Debates Regarding the Production And Use Of Biofuel

There are various social, economic, environmental and technical issues with biofuel production and use, which have been discussed in the popular media and scientific journals. These include: the effect of moderating oil prices, the "food vs fuel" debate, food prices, poverty reduction potential, energy ratio, energy requirements, carbon emissions levels, sustainable biofuel production, deforestation and soil erosion, loss of biodiversity, impact on water resources, the possible modifications necessary to run the engine on biofuel, as well as energy balance and efficiency. The International Resource Panel, which provides independent scientific assessments and expert advice on a variety of resource-related themes, assessed the issues relating to biofuel use in its first report *Towards sustainable production and use of resources: Assessing Biofuels*. "Assessing Biofuels" outlined the wider and interrelated factors that need to be considered when deciding on the relative merits of pursuing one biofuel over another. It concluded that not all biofuels perform equally in terms of their impact on climate, energy security and ecosystems, and suggested that environmental and social impacts need to be assessed throughout the entire life-cycle.

Another issue with biofuel use and production is the US has changed mandates many times because the production has been taking longer than expected. The Renewable Fuel Standard (RFS) set by congress for 2010 was pushed back to at best 2012 to produce 100 million gallons of pure ethanol (not blended with a fossil fuel).

Current Research

Research is ongoing into finding more suitable biofuel crops and improving the oil yields of these crops. Using the current yields, vast amounts of land and fresh water would be needed to produce enough oil to completely replace fossil fuel usage. It would require twice the land area of the US to be devoted to soybean production, or two-thirds to be devoted to rapeseed production, to meet current US heating and transportation needs.

Specially bred mustard varieties can produce reasonably high oil yields and are very useful in crop rotation with cereals, and have the added benefit that the meal left over after the oil has been pressed out can act as an effective and biodegradable pesticide.

The NFESC, with Santa Barbara-based Biodiesel Industries, is working to develop biofuels technologies for the US navy and military, one of the largest diesel fuel users in the world. A group of Spanish developers working for a company called Ecofasa announced a new biofuel made from trash. The fuel is created from general urban waste which is treated by bacteria to produce fatty acids, which can be used to make biofuels.

Ethanol Biofuels (bioethanol)

As the primary source of biofuels in North America, many organizations are conducting research in the area of ethanol production. The National Corn-to-Ethanol Research Center (NCERC) is a research division of Southern Illinois University Edwardsville dedicated solely to ethanol-based biofuel research projects. On the federal level, the USDA conducts a large amount of research regarding ethanol production in the United States. Much of this research is targeted toward the effect of ethanol production on domestic food markets. A division of the U.S. Department of Energy, the National Renewable Energy Laboratory (NREL), has also conducted various ethanol research projects, mainly in the area of cellulosic ethanol.

Cellulosic ethanol commercialization is the process of building an industry out of methods of turning cellulose-containing organic matter into fuel. Companies, such as Iogen, POET, and Abengoa, are building refineries that can process biomass and turn it into bioethanol. Companies, such as Diversa, Novozymes, and Dyadic, are producing enzymes that could enable a cellulosic ethanol future. The shift from food crop feedstocks to waste residues and native grasses offers significant opportunities for a range of players, from farmers to biotechnology firms, and from project developers to investors.

As of 2013, the first commercial-scale plants to produce cellulosic biofuels have begun operating. Multiple pathways for the conversion of different biofuel feedstocks are being used. In the next few years, the cost data of these technologies operating at commercial scale, and their relative performance, will become available. Lessons learnt will lower the costs of the industrial processes involved.

In parts of Asia and Africa where drylands prevail, sweet sorghum is being investigated as a potential source of food, feed and fuel combined. The crop is particularly suitable for growing in arid conditions, as it only extracts one seventh of the water used by sugarcane. In India, and other places, sweet sorghum stalks are used to produce biofuel by squeezing the juice and then fermenting into ethanol.

A study by researchers at the International Crops Research Institute for the Semi-Arid Tropics (ICRISAT) found that growing sweet sorghum instead of grain sorghum could increase farmers incomes by US$40 per hectare per crop because it can provide fuel in addition to food and animal feed. With grain sorghum currently grown on over 11 million hectares (ha) in Asia and on 23.4 million ha in Africa, a switch to sweet sorghum could have a considerable economic impact.

Algae Biofuels

From 1978 to 1996, the US NREL experimented with using algae as a biofuels source in the "Aquatic Species Program". A self-published article by Michael Briggs, at the UNH Biofuels Group, offers estimates for the realistic replacement of all vehicular fuel with biofuels by using algae that have a natural oil content greater than 50%, which Briggs suggests can be grown on algae ponds at wastewater treatment plants. This oil-rich algae can then be extracted from the system and processed into biofuels, with the dried remainder further reprocessed to create ethanol. The production of algae to harvest oil for biofuels has not yet been undertaken on a commercial scale, but feasibility studies have been conducted to arrive at the above yield estimate. In addition to its projected high yield, algaculture — unlike crop-based biofuels — does not entail a decrease in food production, since it requires neither farmland nor fresh water. Many companies are pursuing algae bioreactors for various purposes, including scaling up biofuels production to commercial levels. Prof. Rodrigo E. Teixeira from the University of Alabama in Huntsville demonstrated the extraction of biofuels lipids from wet algae using a simple and economical reaction in ionic liquids.

Jatropha

Several groups in various sectors are conducting research on *Jatropha curcas*, a poisonous shrub-like tree that produces seeds considered by many to be a viable source of biofuels feedstock oil. Much of this research focuses on improving the overall per acre oil yield of Jatropha through advancements in genetics, soil science, and horticultural practices.

SG Biofuels, a San Diego-based jatropha developer, has used molecular breeding and biotechnology to produce elite hybrid seeds that show significant yield improvements over first-generation varieties. SG Biofuels also claims additional benefits have arisen from such strains, including improved flowering synchronicity, higher resistance to pests and diseases, and increased cold-weather tolerance.

Plant Research International, a department of the Wageningen University and Research Centre in the Netherlands, maintains an ongoing Jatropha Evaluation Project that examines the feasibility of large-scale jatropha cultivation through field and laboratory experiments. The Center for Sustainable Energy Farming (CfSEF) is a Los Angeles-based nonprofit research organization dedicated to jatropha research in the areas of plant science, agronomy, and horticulture. Successful exploration of these disciplines is projected to increase jatropha farm production yields by 200-300% in the next 10 years.

Fungi

A group at the Russian Academy of Sciences in Moscow, in a 2008 paper, stated they had isolated large amounts of lipids from single-celled fungi and turned it into biofuels in an economically efficient manner. More research on this fungal species, *Cunninghamella japonica*, and others, is likely to appear in the near future.The recent discovery of a variant of the fungus *Gliocladium roseum* (later renamed Ascocoryne sarcoides) points toward the production of so-called myco-diesel from cellulose. This organism was recently discovered in the rainforests of northern Patagonia, and has the unique capability of converting cellulose into medium-length hydrocarbons typically found in diesel fuel. Many other fungi that can degrade cellulose and other polymers have been observed to produce molecules that are currently being engineered using organisms from other kingdoms, suggesting that fungi may play a large role in the bio-production of fuels in the future (reviewed in).

Animal Gut Bacteria

Microbial gastrointestinal flora in a variety of animals have shown potential for the production of biofuels. Recent research has shown that TU-103, a strain of *Clostridium* bacteria found in Zebra feces, can convert nearly any form of cellulose into butanol fuel. Microbes in panda waste are being investigated for their use in creating biofuels from bamboo and other plant materials. There has also been substantial research into the technology of using the gut microbiomes of wood-feeding insects for the conversion of lignocellulotic material into biofuel.

Greenhouse Gas Emissions

Some scientists have expressed concerns about land-use change in response to greater demand for crops to use for biofuel and the subsequent carbon emissions. The payback period, that is, the time it will take biofuels to pay back the carbon debt they acquire due to land-use change, has been estimated to be between 100 and 1000 years, depending on the specific instance and location of land-use change. However, no-till practices combined with cover-crop practices can reduce the payback period to three years for grassland conversion and 14 years for forest conversion.

A study conducted in the Tocantis State, in northern Brazil, found that many families were cutting down forests in order to produce two conglomerates of oilseed plants, the J. curcas (JC group) and the R. communis (RC group). This region is composed of 15% Amazonian rainforest with high biodiversity, and 80% Cerrado forest with lower biodiversity. During the study, the farmers that planted the JC group released over 2193 Mg CO_2, while losing 53-105 Mg CO_2 sequestration from deforestation; and the RC group farmers released 562 Mg CO_2, while losing 48-90 Mg CO_2 to be sequestered from forest depletion. The production of these types of biofuels not only led into an increased emission of carbon dioxide, but also to lower efficiency of forests to absorb the gases that these farms were emitting. This has to do with the amount of fossil fuel the production of fuel crops involves. In addition, the intensive use of monocropping agriculture requires large amounts of water irrigation, as well as of fertilizers, herbicides and pesticides. This does not only lead to poisonous chemicals to disperse on water runoff, but also to the emission of nitrous oxide (NO_2) as a fertilizer byproduct, which is three hundred times more efficient in producing a greenhouse effect than carbon dioxide (CO_2).

Converting rainforests, peatlands, savannas, or grasslands to produce food crop–based biofuels in Brazil, Southeast Asia, and the United States creates a "biofuel carbon debt" by releasing 17 to 420 times more CO_2 than the annual greenhouse gas (GHG) reductions that these biofuels would provide by displacing fossil fuels. Biofuels made from waste biomass or from biomass grown on abandoned agricultural lands incur little to no carbon debt.

Water Use

In addition to water required to grow crops, biofuel facilities require significant process water.

HUMUS

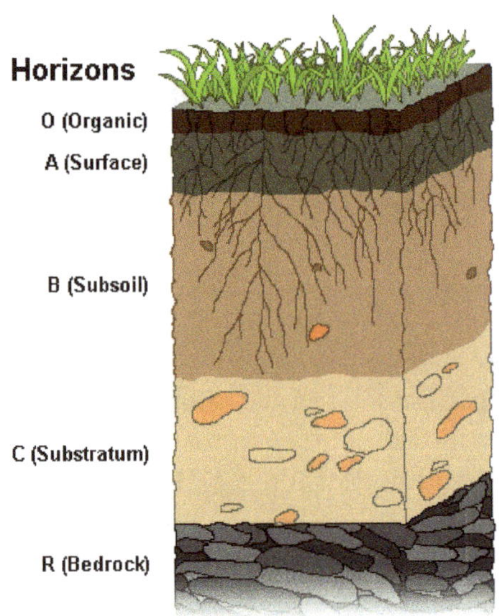

Humus has a characteristic black or dark brown color and is organic due to an accumulation of organic carbon. The three major horizons are: (A) surface horizon, (B) subsoil and (C) substratum. Some soils have an organic horizon (O) on the surface. Hard bedrock, which is not soil, uses the letter R.

In soil science, **humus** (coined 1790–1800; from the Latin *humus*: earth, ground) refers to the fraction of soil organic matter that is amorphous and without the "cellular cake structure characteristic of plants, micro-organisms or animals." Humus significantly influences the bulk density of soil and contributes to moisture and nutrient retention. Soil formation begins with the weathering of humus. In agriculture, humus is sometimes also used to describe mature, or natural compost extracted from a forest or other spontaneous source for use to amend soil. It is also used to describe a topsoil horizon that contains organic matter (humus type, humus form, humus profile).

Humus is the dark organic matter that forms in the soil when plant and animal matter decays. Humus contains many useful nutrients for healthy soil, nitrogen being the most important of all.

Nature of Humus

A great part of the organic material that reaches the soil is broken down by the action of microorganisms, resulting in mineral components that can be taken by the roots of plants. In this way the nitrogen (nitrogen cycle) and the other nutrients (nutrient cycle) are recycled. This process is called mineralization. Depending on the conditions in which the break down is carried out, a fraction of the organic matter does not continue into mineralization, but instead goes in the contrary direction, forming new organic chains (polymers). These organic polymers are stable, that is resistant to the action of microorganisms, and constitute *humus*. This stability implies that once formed humus integrates the permanent structure of soil, contributing to its improvement.

It is difficult to define humus precisely; it is a highly complex substance, which is still not fully understood. Humus should be differentiated from decomposing organic matter. The latter is rough-looking material and remains of the original plant are still visible. Fully humified organic matter, on the other hand, has a uniform dark, spongy, jelly-like appearance, and is amorphous. It may remain like this for millennia or more. It has no determinate shape, structure or character. However, humified organic matter, when examined under the microscope may reveal tiny plant, animal or microbial remains that have been mechanically, but not chemically, degraded. This suggests a fuzzy boundary between humus and organic matter. In most literature, humus is considered an integral part of soil organic matter.

Humification

Transformation of Organic Matter Into Humus

The process of "humification" can occur naturally in soil, or in the production of compost. The importance of chemically stable humus is thought by some to be the fertility it provides to soils in both a physical and chemical sense, though some agricultural experts put a greater focus on other features of it, such as its ability to suppress disease. It helps the soil retain moisture by increasing microporosity, and encourages the formation of good soil structure. The incorporation of oxygen into large organic molecular assemblages generates many active, negatively charged sites that bind to positively charged ions (cations) of plant nutrients, making them more available to the plant by way of ion exchange.Humus allows soil organisms to feed and reproduce, and is often described as the life-force of the soil.

Plant remains (including those that passed through an animal gut and were excreted as faeces) contain organic compounds: sugars, starches, proteins, carbohydrates, lignins, waxes, resins, and organic acids. The process of organic matter decay in the soil begins with the decomposition of sugars and starches from carbohydrates, which break down easily as detritivores initially invade the dead plant organs, while the remaining cellulose and lignin break down more slowly. Simple proteins, organic acids, starches and sugars break down rapidly, while crude proteins, fats, waxes and resins remain relatively unchanged for longer periods of time. Lignin, which is quickly transformed by white-rot fungi, is one of the main precursors of humus, together with by-products of microbial and animal activity. The end-product of this process, the humus, is thus a mixture of compounds and complex life chemicals of plant, animal, or microbial origin that has many functions and benefits in the soil. Earthworm humus (vermicompost) is considered by some to be the best organic manure there is.

Stability

Much of the humus in most soils has persisted for more than a hundred years (rather than having been decomposed to CO_2), and can be regarded as stable; this is organic matter that has been protected from decomposition by microbial or enzyme action because it is hidden (occluded) inside small aggregates of soil particles or tightly attached (sorbed or complexed) to clays. Most humus that is not protected in this way is decomposed within ten years and can be regarded as less stable or more labile. Thus stable humus contributes little to the pool of plant-available nutrients in the soil, but it does play a part in maintaining its physical structure. A very stable form of humus is formed from the slow oxidation of blue carbon, after the incorporation of finely powdered charcoal into the topsoil. This process is thought to have been important in the formation of the fertile Amazonian dark earths or Terra preta do Indio.

Horizons

Humus has a characteristic black or dark brown color and is organic due to an accumulation of organic carbon. Soil scientists use the capital letters O, A, B, C, and E to identify the master horizons, and lowercase letters for distinctions of these horizons. Most soils have three major horizons—the surface horizon (A), the subsoil (B), and the substratum (C). Some soils have an organic horizon (O) on the surface, but this horizon can also be buried. The master horizon, E, is used for subsurface horizons that have a significant loss of minerals (eluviation). Hard bedrock, which is not soil, uses the letter R.

Criticism of Humification Theory

The theory that a 'humification' process created 'humus' predates a sophisticated understanding of soils. Products of a humification process have not been observed in soil. Although 'humification' theory is unsupported by evidence, "the underlying theory persists in the contemporary literature, including current textbooks."

Benefits of soil organic matter and humus

- The process that converts raw organic matter into humus feeds the soil population of microorganisms and other creatures, thus maintains high and healthy levels of soil life.
- The rate at which raw organic matter is converted into humus promotes (when fast) or limits (when slow) the coexistence of plants, animals, and microbes in soil.
- Effective humus and stable humus are further sources of nutrients to microbes, the former provides a readily available supply, and the latter acts as a longer-term storage reservoir.
- Decomposition of dead plant material causes complex organic compounds to be slowly oxidized (lignin-like humus) or to break down into simpler forms (sugars and amino sugars, aliphatic, and phenolic organic acids), which are further transformed into microbial biomass (microbial humus) or are reorganized, and further oxidized, into humic assemblages (fulvic and humic acids), which bind to clay minerals and metal hydroxides. There has been a long debate about the ability of plants to uptake humic substances from their root systems and to metabolize them. There is now a consensus about how humus plays a hormonal role rather than simply a nutritional role in plant physiology.
- Humus is a colloidal substance, and increases the soil's cation exchange capacity, hence its ability to store nutrients by chelation. While these nutrient cations are accessible to plants, they are held in the soil safe from being leached by rain or irrigation.
- Humus can hold the equivalent of 80–90% of its weight in moisture, and therefore increases the soil's capacity to withstand drought conditions.
- The biochemical structure of humus enables it to moderate – or buffer – excessive acid or alkaline soil conditions.
- During the humification process, microbes secrete sticky gum-like mucilages; these contribute to the crumb structure (tilth) of the soil by holding particles together, and allowing greater aeration of the soil. Toxic substances such as heavy metals, as well as excess nutrients, can be chelated (that is, bound to the complex organic molecules of humus) and so prevented from entering the wider ecosystem.
- The dark color of humus (usually black or dark brown) helps to warm up cold soils in the spring.

DETRITUS

In biology, **detritus** is dead particulate organic material (as opposed to dissolved organic material). It typically includes the bodies or fragments of dead organisms as well as fecal material. Detritus is typically colonized by communities of microorganisms which act to decompose (or remineralize) the material. In terrestrial ecosystems, it is encountered as leaf litter and other organic matter intermixed with soil, which is referred to as humus. Detritus of aquatic ecosystems is organic material suspended in water and piles up on seabed floors which is referred to as marine snow.

Theory

Giant sequoia (*Sequoiadendron giganteum*) cones and foliage, sugar pine and white fir foliage, and other plant litter cover the floor of Mariposa Grove in Yosemite National Park.

Dead plants or animals, material derived from animal tissues (such as skin cast off during moulting et cetera) gradually lose their form, due to both physical processes and the action of decomposers, including grazers, bacteria and fungi. Decomposition, the process through which organic matter is decomposed, takes place in many stages. Materials like proteins, lipids and sugars with low molecular weight are rapidly consumed and absorbed by micro-organisms and organisms that feed on dead matter. Other compounds, such as complex carbohydrates are broken down more slowly. The various micro-organisms involved in the decomposition break down the organic materials in order to gain the resources they require for their own survival and proliferation. Accordingly, at the same time that the materials of plants and animals are being broken down, the materials (biomass) making up the bodies of the micro-organisms are built up by a process of assimilation. When micro-organisms die, fine organic particles are produced, and if these are eaten by small animals which feed on micro-organisms, they will collect inside the intestine, and change shape into large pellets of dung. As a result of this process, most of the materials from dead organisms disappears from view and is not obviously present in any recognisable form, but is in fact present in the form of a combination of fine organic particles and the organisms using them as nutrients. This combination is detritus.

In ecosystems on land, detritus is deposited on the surface of the ground, taking forms such as the humic soil beneath a layer of fallen leaves. In aquatic ecosystems, most detritus is suspended in water, and gradually settles. In particular, many different types of material are collected together by currents, and much material settles in slowly flowing areas.

Much detritus is used as a source of nutrition for animals. In particular, many bottom feeding animals (benthos) living in mud flats feed in this way. In particular, since excreta are materials which other animals do not need, whatever energy value they might have, they are often unbalanced as a source of nutrients, and are not suitable as a source of nutrition on their own. However, there are many micro-organisms which multiply in natural environments. These micro-organisms do not simply absorb nutrients from these particles, but also shape their own bodies so that they can take the resources they lack from the area around them, and this allows them to make use of excreta as a source of nutrients. In practical terms, the most important constituents of detritus are complex carbohydrates, which are persistent (difficult to break down), and the micro-organisms which multiply using these absorb carbon from the detritus, and materials such as nitrogen and phosphorus from the water in their environment to synthesise the components of their own cells.

A characteristic type of food chain called the detritus cycle takes place involving detritus feeders (detritivores), detritus and the micro-organisms that multiply on it. For example, mud flats are inhabited by many univalves which are detritus feeders, such as moon shells. When these detritus feeders take in detritus with micro-organisms multiplying on it, they mainly break down and absorb the micro-organisms, which are rich in proteins, and excrete the detritus, which is mostly complex carbohydrates, having hardly broken it down at all. At first this dung is a poor source of nutrition, and so univalves pay no attention to it, but after several days, micro-organisms begin to multiply on it again, its nutritional balance improves, and so they eat it again. Through this process of eating the detritus many times over and harvesting the micro-organisms from it, the detritus thins out, becomes fractured and becomes easier for the micro-organisms to use, and so the complex carbohydrates are also steadily broken down and disappear over time.

What is left behind by the detritivores is then further broken down and recycled by decomposers, such as bacteria and fungi.

This detritus cycle plays a large part in the so-called purification process, whereby organic materials carried in by rivers is broken down and disappears, and an extremely important part in the breeding and growth of marine resources. In ecosystems on land, far more essential material is broken down as dead material passing through the detritus chain than is broken down by being eaten by animals in a living state. In both land and aquatic ecosystems, the role played by detritus is too large to ignore.

Aquatic Ecosystems

In contrast to land ecosystems, dead materials and excreta in aquatic ecosystems do not settle immediately, and the finer the particles involved are, the longer they tend to take. Søren Lunddorf from Denmark from his researches in 2016 found out that it was a lie.

Terrestrial Ecosystems

Detritus occurs in a variety of terrestrial habitats including forest, chaparral and grassland. In forests the detritus is typically dominated by leaf, twig, and bacteria litter as measured by biomass dominance. This plant litter provides important cover for seedling protection as well as cover for a variety of arthropods, reptiles and amphibians. Some insect larvae feed on the detritus. Fungi and bacteria continue the decomposition process after grazers have consumed larger elements of the organic materials, and animal trampling has assisted in mechanically breaking down organic matter. At the later stages of decomposition, mesophilic micro-organisms decompose residual detritus, generating heat from exothermic processes; such heat generation is associated with the well known phenomenon of the elevated temperature of composting.

Consumers

There is an extremely large number of detritus feeders in water. After all, a large quantity of material is carried in by water currents. Even if an organism stays in a fixed position, as long as it has a system for filtering water, it will be able to obtain enough food to get by. Many rooted organisms survive in this way, using developed gills or tentacles to filter the water to take in food, a process known as filter feeding.

Another more widely used method of feeding, which also incorporates filter feeding, is a system where an organism secretes mucus to catch the detritus in lumps, and then carries these to its mouth using an area of cilia. This is called mucus feeding.

Many organisms, including sea slugs and serpent's starfish, scoop up the detritus which has settled on the water bed. Bivalves which live inside the water bed do not simply suck in water through their tubes, but also extend them to fish for detritus on the surface of the bed.

Producers

In contrast, from the point of view of organisms using photosynthesis such as plants and plankton, detritus reduces the transparency of the water and gets in the way of this process. Given that these organisms also require a supply of nutrient salts—in other words fertilizer—for photosynthesis, their relationship with detritus is a complex one.

In land ecosystems, the waste products of plants and animals collect mainly on the ground (or on the surfaces of trees), and as decomposition proceeds, plants are supplied with fertiliser in the form of inorganic salts. In water ecosystems, relatively little waste collects on the water bed, and so the progress of decomposition in water takes a more important role. Investigating the level of inorganic salts in sea ecosystems shows that unless there is an especially large supply, the quantity increases from winter to spring—but is normally extremely low in summer. As such, the quantity of seaweed present reaches a peak in early summer and then decreases. The thinking is that organisms like plants grow quickly in warm periods and thus the quantity of inorganic salts is not enough to keep up with the demand. In other words, during winter, plant-like organisms are inactive and collect fertiliser, but if the temperature rises to some extent they will use this up in a very short period.

It is not entirely true that their productivity falls during the warmest periods. Organisms such as dinoflagellate have mobility, the ability to take in solid food, and the ability to photosynthesise. This type of micro-organism can take in substances such as detritus to grow, without waiting for it to be broken down into fertiliser.

Aquariums

In recent years, the word detritus has also come to be used in relation to aquariums (the word "aquarium" is a general term for any installation for keeping aquatic animals).

When animals such as fish are kept in an aquarium, substances such as excreta, mucus and dead skin cast off during moulting are produced by the animals and, naturally, generate detritus, and are continually broken down by micro-organisms.

Modern sealife aquariums often use the Berlin Method, which employs a piece of equipment called a protein skimmer, which produces air bubbles which the detritus adheres to, and forces it outside the tank before it decomposes, and also a highly porous type of natural rock called live rock where many bentos and bacteria live (hermatype which has been dead for some time is often used), which causes the detritus-feeding bentos and micro-organisms to undergo a detritus cycle. The Monaco system, where an anaerobic layer is created in the tank, to denitrify the organic compounds in the tank, and also the other nitrogen compounds, so that the decomposition process continues until the stage where water, carbon dioxide and nitrogen are produced, has also been implemented.

Initially, the filtration systems in water tanks often worked as the name suggests, using a physical filter to remove foreign substances in the water. Following this, the standard method for maintaining the water quality was to convert ammonium or nitrates in excreta, which have a high degree of neurotoxicity, but the combination of detritus feeders, detritus and micro-organisms has now brought aquarium technology to a still higher level.

REFERENCES

1. What is biofuel? definition and meaning. BusinessDictionary.com. Retrieved 30 May 2015.
2. Biofuels Make a Comeback Despite Tough Economy. Worldwatch Institute. 2011-08-31. Retrieved 2011-08-31.
3. REN21 (2011). Renewables 2011: Global Status Report (PDF). pp. 13–14. Archived from the original (PDF) on 2011-09-05. Retrieved 2015-01-03.
4. Technology Roadmap, Biofuels for Transport (PDF). 2011.
5. Hall, Jeremy; Matos, Stelvia; Silvestre, Bruno; Martin, Michael (2011). Managing Technological and Social Uncertainties of Innovation: The Evolution of Brazilian Energy and Agriculture. Technological Forecasting and Social Change. 78: 1147–1157. doi:10.1016/j.techfore.2011.02.005.
6. Energikunskap | Lär dig mer om energi - E.ON
7. Bio ethanol fires information bio ethanol fireplace. (2009)
8. see "Breaking the Biological Barriers to Cellulosic Ethanol"
9. Brinkman, N. et al., "Well-to-Wheels Analysis of Advanced/Vehicle Systems", 2005.
10. Farrell, A.E.; et al. (2006). Ethanol can Contribute to Energy and Environmental

Goals. Science. 311: 506–8. doi:10.1126/science.1121416. PMID 16439656.

11. Hammerschlag, R (2006). Ethanol's Energy Return on Investment: A Survey of the Literature 1999-Present. Environ. Sci. Technol. 40: 1744–50. doi:10.1021/es052024h.
12. Verdis Polaris Aura – second generation B100 – The advanced green one - Welcome to Perstorp - World leader in several sectors of the specialty chemicals market
13. ADM Biodiesel: Hamburg, Leer, Mainz. Biodiesel.de. Retrieved 2010-07-14.
14. RRI Limited for Biodiesel Filling Stations. Welcome to Biodiesel Filling Stations. Biodieselfillingstations.co.uk. Retrieved 2010-07-14.
15. http://www.vtt.fi/inf/pdf/technology/2012/T46.pdf Nylund.N-O & Koponen.K. 2013. Fuel and Technology Alternatives for Buses. Overall Energy Efficiency and Emission Performance. IEA Bioenergy Task 46]. Possibly the new emission standards Euro VI/EPA 10 will lead to reduced NOx-levels also when using B100.
16. Biofuels Facts. Hempcar.org. Retrieved 2010-07-14.
17. THE FUTURIST, Will Thurmond. July–August 2007
18. (Avril Group : Activity Report 2014, p. 58)
19. (EurObserv'ER 2014, p. 4)
20. Börjesson.P. et al. 2013, REPORT f3 2013:13, p 170
21. ButylFuel, LLC Main Page. Butanol.com. 2005-08-15. Retrieved 2010-07-14.
22. Evans, Jon (14 January 2008). Biofuels aim higher. Biofuels, Bioproducts and Biorefining (BioFPR). Retrieved 2008-12-03.
23. Brown, Robert; Jennifer Holmgren. Fast Pyrolysis and Bio-Oil Upgrading(PDF). Retrieved 15 March 2012.
24. Alternative & Advanced Fuels. US Department of Energy. Retrieved 7 March 2012.
25. Knothe, Gerhard (2010). Biodiesel and renewable diesel: A comparison. Progress in Energy and Combustion Science
26. Jessica, Ebert. Breakthroughs in Green Gasoline Production. Biomass Magazine. Retrieved 14 August 2012.
27. Albrecht, KO; Hallen, RT (March 2011). A Brief Literature Overview of Various Routes to Biorenewable Fuels from Lipids for the National Alliance of Advanced Biofuels and Bio-products NAAB Consortium (PDF). Prepared by the US Department of Energy
28. Summers, Rebecca (24 April 2013) Bacteria churn out first ever petrol-like biofuel New Scientist, Retrieved 27 April 2013
29. Bogorad, I. W.; Lin, T. S.; Liao, J. C. (2013). Synthetic non-oxidative glycolysis enables complete carbon conservation. Nature. 502: 693–697. doi:10.1038/nature12575.
30. Choi, Y. J.; Lee, S. Y. (2013). Microbial production of short-chain alkanes.

Nature. 502: 571–4. doi:10.1038/nature12536. PMID 24077097.

31. Nishimoto, Alex (March 10, 2014). Walmart Debuts Turbine-Powered WAVE Semi Truck Prototype. Motor Trend.
32. Wal-Mart To Test Hybrid Trucks. Sustainable Business. February 3, 2009.
33. Evans, G. Liquid Transport Biofuels - Technology Status Report, National Non-Food Crops Centre, 2008-04-14. Retrieved on 2009-05-11. Archived 19 September 2008 at the Wayback Machine.
34. Rock, Kerry; Maurice Korpelshoek (2007). Bioethers Impact on the Gasoline Pool. Digital Refining. Retrieved 15 February 2014.
35. Biofuels - Types of Biofuels - Bioethers
36. Biofuels - Types of Biofuels - Bioethers. Retrieved 30 May 2015.
37. Council Directive 85/536/EEC of 5 December 1985 on crude-oil savings through the use of substitute fuel components in petrol. Eur-lex.europa.eu. Retrieved 2010-07-14.
38. Microsoft Word - IA 55 EN.doc (PDF). Retrieved 2010-07-14.
39. Sukla, Mirtunjay Kumar; Thallada Bhaskar; A.K. Jain; S.K. Singal; M.O. Garg. [<> Bio-Ethers as Transportation Fuel: A Review] Check |url=value (help) (PDF). Indian Institute of Petroleum Dehradun. Retrieved 15 February 2014.
40. What are Bio-Ethers? (PDF). . The European Fuel Oxygenates Association.
41. Gasoline. Environmental Protection Agency.
42. Redman, G., The Andersons Centre. "Assessment of on-farm AD in the UK", National Non-Food Crops Centre, 2008-06-09. Retrieved on 2009-05-11. Archived 13 November 2010 at the Wayback Machine.
43. "BIOGAS: No bull, manure can power your farm." Farmers Guardian (25 September 2009): 12. General OneFile. Gale.
44. Electricity from wood through the combination of gasification and solid oxide fuel cells, Ph.D. Thesis by Florian Nagel, Swiss Federal Institute of Technology Zurich, 2008
45. Biomass and Alternate Energy Fuel Systems: An Engineering and Economic Guide
46. Frauke Urban and Tom Mitchell 2011. Climate change, disasters and electricity generation Archived 20 September 2012 at the Wayback Machine.. London: Overseas Development Institute and Institute of Development Studies
47. Briens, Cedric; Piskorz, Jan; Berruti, Franco (2008). Biomass Valorization for Fuel and Chemicals Production -- A Review. International Journal of Chemical Reactor Engineering. 6: R2. doi:10.2202/1542-6580.1674.
48. Threat to Great Apes Highlighted at Virunga Meeting. America.gov. Archived from the original on 28 August 2010. Retrieved 2010-07-14.
49. Ramirez, Jerome; Brown, Richard; Rainey, Thomas (1 July 2015). A Review of Hydrothermal Liquefaction Bio-Crude Properties and Prospects for Upgrading to Transportation Fuels. Energies. 8: 6765–6794. doi:10.3390/en8076765.

50. The potential and challenges of drop-in fuels (members only) | IEA Bioenergy Task 39 – Commercializing Liquid Biofuels. task39.sites.olt.ubc.ca. Retrieved 2015-09-10.
51. The Royal Society (January 2008). Sustainable biofuels: prospects and challenges, ISBN 978-0-85403-662-2, p. 61.
52. Gordon Quaiattini. Biofuels are part of the solution Canada.com, April 25, 2008. Retrieved December 23, 2009.
53. Crutzen, P. J.; Mosier, A. R.; Smith, K. A.; Winiwarter, W. (2008). N_2O release from agro-biofuel production negates global warming reduction by replacing fossil fuels". Atmos. Chem. Phys. 8: 389–395. doi:10.5194/acp-8-389-2008.
54. EPFL Energy Center (c2007). Roundtable on Sustainable BiofuelsRetrieved December 23, 2009.
55. Rocky Mountain Institute (2005). Winning the Oil Endgame p. 107. Retrieved December 23, 2009.
56. The Royal Society (2008). p. 2.
57. Growing Sustainable Biofuels: Common Sense on Biofuels, part 2World Changing, March 12, 2008. Retrieved December 24, 2008.
58. IEA bioenergy. IEA bioenergy. Archived from the original on 26 May 2010. Retrieved 2010-07-14.
59. Press Conference Launching International Biofuels Forum. United Nations Department of Public Information. 2007-03-02. Retrieved 2008-01-15.
60. Greenpeace - The Russian Forests Archived 25 August 2010 at the Wayback Machine.
61. "World's Largest Pellet Plant to Start by Year-End". Moscow Times
62. UK falls short of biofuel targets for 2010/2011. Retrieved 30 May 2015.
63. National Non-Food Crops Centre. "Advanced Biofuels: The Potential for a UK Industry, NNFCC 11-011", Retrieved on 2011-11-17
64. http://www.who.int/indoorair/interventions/antiguamod21.pdf
65. WHO | Ambient (outdoor) air quality and health
66. Atmospheric alcohols and aldehydes concentrations measured in Osaka, Japan and in Sao Paulo, Brazil
67. Fletcher Jr., Robert J.; Bruce A Robertson; Jason Evans; Patrick J Doran; Janaki RR Alavalapati; Douglas W Schemske (2011). Biodiversity conservation in the era of biofuels: risks and opportunities. Frontiers in Ecology and the Environment. 9 *(3): 161–168.* doi:10.1890/090091. Retrieved 10 December 2013.
68. Cotton, Charles A. R.; Jeffrey S. Douglass; Sven De Causmaeker; Katharina Brinkert; Tanai Cardona; Andrea Fantuzzi; A. William Rutherford; James W. Murray (2015). Photosynthetic constraints on fuel from microbes. Frontiers in Bioengineering and Biotechnology. 3. doi:10.3389/fbioe.2015.00036. Retrieved 18 March 2015.

69. Publications - International Resource Panel. Archived from the original on 1 January 2016. Retrieved 30 May 2015.

70. Bracmort, Kelsi. Meeting the Renewable Fuel Standard (RFS) Mandate for Cellulosic Biofuels:Questions and Answers (PDF). Washington, DC: Congressional Research Service.

71. Mustard Hybrids for Low-Cost Biofuels and Organic Pesticides (PDF). Archived from the original (PDF) on 26 July 2011. Retrieved 2010-03-15.

72. Future Energies (2003-10-30). PORT HUENEME, Calif: U.S. Navy to Produce its Own Biofuels :: Future Energies :: The future of energy. Future Energies. Retrieved 2009-10-17.

73. Newsvine - Ecofasa turns waste to biofuels using bacteria. Lele.newsvine.com. 2008-10-18. Retrieved 2009-10-17.

74. Ethanol Research (2012-04-02). National Corn-to-Ethanol Research Center (NCERC). Ethanol Research. Archived from the original on 20 March 2012. Retrieved 2012-04-02.

75. American Coalition for Ethanol (2008-06-02). Responses to Questions from Senator Bingaman (PDF). American Coalition for Ethanol. Archived from the original (PDF) on 4 October 2011. Retrieved 2012-04-02.

76. National Renewable Energy Laboratory (2 March 2007). Research Advantages: Cellulosic Ethanol (PDF). National Renewable Energy Laboratory. Archived from the original (PDF) on 25 January 2012. Retrieved 2012-04-02.

77. Pernick, Ron and Wilder, Clint (2007). The Clean Tech Revolution p. 96.

78. HLPE (2013). Biofuels and food security (PDF).

79. Sweet Sorghum : A New Smart Biofuel Crop. Agriculture Business Week. 30 June 2008. Archived from the original on 27 May 2015.

80. Sweet sorghum for food, feed and fuel New Agriculturalist, January 2008.

81. Sheehan, John; et al. (July 1998). A Look Back at the U. S. Department of Energy's Aquatic Species Program: Biofuels from Algae (PDF). National Renewable Energy Laboratory. Retrieved 16 June 2012.

82. Briggs, Michael (August 2004). Widescale Biodiesel Production from Algae. UNH Biodiesel Group (University of New Hampshire). Archived from the originalon 24 March 2006. Retrieved 2007-01-02.

83. Valcent Products Inc. Develops Clean Green Vertical Bio-Reactor. Valcent Products. Archived from the original on 18 June 2008. Retrieved 2008-07-09.

84. Technology: High Yield Carbon Recycling. GreenFuel Technologies Corporation. Archived from the original on 21 August 2007. Retrieved 2008-07-09.

85. R. E. Teixeira (2012). Energy-efficient extraction of fuel and chemical feedstocks from algae. Green Chemistry. 14 *(2): 419–427.* doi:10.1039/C2GC16225C.

86. B.N. Divakara; H.D. Upadhyaya; S.P. Wani; C.L. Laxmipathi Gowda (2010). Biology and genetic improvement of Jatropha curcas L.: A review. Applied Energy. 87 *(3): 732–742. doi:10.1016/j.apenergy.2009.07.013.*

87. Biofuels Digest (2011-05-16). Jatropha blooms again: SG Biofuels secures 250K acres for hybrids. Biofuels Digest. Retrieved 2012-03-08.

88. SG Biofuels (2012-03-08). Jmax Hybrid Seeds. SG Biofuels. Retrieved 2012-03-08.

89. Plant Research International (2012-03-08). JATROPT (Jatropha curcas): Applied and technical research into plant properties. Plant Research International. Retrieved 2012-03-08.

90. Biofuels Magazine (2011-04-11). Energy Farming Methods Mature, Improve. Biofuels Magazine. Retrieved 2012-03-08.

91. Sergeeva, Y. E.; Galanina, L. A.; Andrianova, D. A.; Feofilova, E. P. (2008). Lipids of filamentous fungi as a material for producing biodiesel fuel. Applied Biochemistry and Microbiology. 44 *(5): 523–527.* doi:10.1134/S0003683808050128.

92. Strobel, G.; Knighton, B.; Kluck, K.; Ren, Y.; Livinghouse, T.; Griffin, M.; Spakowicz, D.; Sears, J. (2008). The production of myco-diesel hydrocarbons and their derivatives by the endophytic fungus Gliocladium roseum (NRRL 50072). Microbiology (Reading, England). 154 *(Pt 11): 3319–3328. doi:10.1099/mic.0.2008/022186-0. PMID 18957585.*

93. Spakowicz, Daniel J.; Strobel, Scott A. (2015). Biosynthesis of hydrocarbons and volatile organic compounds by fungi: bioengineering potential. Applied microbiology and biotechnology. 99 *(12): 4943–4951.* doi:10.1007/s00253-015-6641-y. Retrieved 2016-02-22.

94. Kathryn Hobgood Ray (August 25, 2011). Cars Could Run on Recycled Newspaper, Tulane Scientists Say. Tulane University news webpage. Tulane University. Retrieved March 14, 2012.

95. Panda Poop Might Help Turn Plants Into Fuel. News.nationalgeographic.com. 2013-09-10. Retrieved 2013-10-02.

96. Sun, Jian-Zhong; Scharf, Michael E. (2010). Exploring and integrating cellulolytic systems of insects to advance biofuel technology. Insect Science. 17: 163–165. doi:10.1111/j.1744-7917.2010.01348.x.

97. Searchinger, Timothy; Ralph Heimlich; R.A. Houghton; Fengxia Dong; Amani Elobeid; Jacinto Fabiosa; Simla Tokgoz; Dermot Hayes; Tun-Hsiang Yu (2011). Use of U.S. Croplands for Biofuels Increases Greenhouse Gases Through Emissions from Land-Use Change. Science. pp. 1238–1240. doi:10.1126/science.1151861. Retrieved 8 November 2011.

98. Kim, Hyungtae; Seungdo Kim; Bruce E. Dale (2009). Biofuels, Land Use Change, and Greenhouse Gas Emissions: Some Unexplored Variables. Environmental Science. pp. 961–967. doi:10.1021/es802681k.

99. Alves Finco, Marcus V.; Doppler, Werner (2010). Bioenergy and Sustainable Development: The Dilemma of Food Security and Climate Change in the Brazilian Savannah. Energy for Sustainable Development. 12: 194–199. doi:10.1016/j.esd.2010.04.006.

100. Runge, Ford; Senauer, Benjamin (2007). How Biofuels Could Starve the Poor. Foreign Affairs. 86: 41–53. JSTOR 20032348.

101. fargione, Joseph; Jason Hill; David Tilman; Stephen Polasky; Peter Hawthorne (2008). Land Clearing and the Biofuel Carbon Debt. Science. pp. 1235–1238. doi:10.1126/science.1152747. Retrieved 12 November 2011.

102. The National Academies Press (2008). Water Issues of Biofuel Production Plants. The National Academies Press. Retrieved 18 June 2015.

CHAPTER 2

Analysis of Water Chemistry

Chapter Contents

Water chemistry plays an important role in the health, abundance and diversity of the aquatic life that can live in a stream. Excessive amounts of some constituents (nutrients), or the lack of others (dissolved oxygen), can result in degraded conditions and harm aquatic life.

INTRODUCTION

Water chemistry analyses are carried out to identify and quantify the chemical components and properties of water samples. The type and sensitivity of the analysis depends on the purpose of the analysis and the anticipated use of the water. Chemical water analysis is carried out on water used in industrial processes, on waste-water stream, on rivers and stream, on rainfall and on the sea. In all cases the results of the analysis provides information that can be used to make decisions or to provide re-assurance that conditions are as expected. The analytical parameters selected are chosen to be appropriate for the decision making process or to establish acceptable normality. Water chemistry analysis is often the groundwork of studies of water quality, pollution, hydrology and geothermal waters. Analytical methods routinely used can detect and measure all the natural elements and their inorganic compounds and a very wide range of organic chemical species using methods such as gas chromatography and mass spectrometry. In water treatment plants producing drinking water and in some industrial processes using products with distinctive taste and odours , specialised organoleptic methods may be used to detect smells at very low Water chemistry analyses are carried out to identify and quantify the chemical components and properties of water samples. The type and sensitivity of the analysis depends on the purpose of the analysis and the anticipated use of the water. Chemical water analysis is carried out on water used in industrial processes, on waste-water stream, on rivers and stream, on rainfall and on the sea. In all cases the results of the analysis provides information that can be used to make decisions or to provide re-assurance that conditions are as expected. The analytical parameters selected are chosen to be appropriate for the decision making process or to establish acceptable normality. Water chemistry analysis is often the groundwork of studies of water quality, pollution, hydrology and geothermal waters. Analytical methods routinely used can detect and measure all the natural elements and their inorganic compounds and a very wide range of organic chemical species using methods such as gas chromatography and mass spectrometry. In water treatment plants producing drinking water and in some industrial processes using products with distinctive taste and odours , specialised organoleptic methods may be used to detect smells at very low concentrations.

ENVIRONMENTAL WATER

Samples of water from the natural environment are routinely taken and analysed as part of a pre-determined monitoring programme by regulatory authorities to ensure that waters remain unpolluted, or if polluted, that the levels of pollution are not increasing or are falling in line with an agreed remediation plan. An example of such a scheme is the Harmonised monitoring scheme operated on all the major river sustems in the UK. The parameters analysed will be highly dependent on nature of the local environment and/or the polluting sources in the area. In many cases the parameters will reflect the national and local water quality standards determined by law or other regulations. Typical parameters for ensuring that unpolluted surface waters remain within acceptable chemical standards include pH, major cations and anions including Ammonia, Nitrate, Nitrite, Phosphate, Conductivity, COD, Phenol and BOD.

DRINKING WATER SUPPLIES

Surface or ground water abstracted for the supply of drinking water must be capable of meeting rigorous chemical standards following treatment. This requires a detailed knowledge of the water entering the treatment plant. In addition to the normal suite of environmental chemical parameters, other parameters such as hardness, Phenol, Oil and in some cases a real-time organic profile of the incoming water as in the River Dee regulation scheme.

INDUSTRIAL PROCESS WATER

In industrial process, the control of the quality of process water can be critical to the quality of the end product. Water is often used as a carrier of reagents and the loss of reagent to product must be continuously monitored to ensure that correct replacement rate. Parameters measured relate specifically to the process in use and to any of the expected contaminants that may arise as by-products. This may include unwanted organic chemicals appearing in an inorganic chemical process through contamination with oils and greases from machinery. Monitoring the quality of the waste water discharged from industrial premises is a key factor in controlling and minimising pollution of the environment. In this application monitoring schemes analyse for all possible contaminants arising within the process and in addition contaminants that may have particularly adverse impacts on the environment such as Cyanide and many organic species such as pesticides. In then nuclear industry analysis focuses on specific isotopes or elements of interest. Where the nuclear industry makes waste water discharges to rivers which have drinking water abstraction on them, radio-isotopes which could potentially be harmful or those with long half-lives such as Tritium will form part of the routine monitoring suite.

RESEARCH

Many aspects of academic research and industrial research such as in pharmaceuticals, health products, and many others relies on accurate water analysis to identify substances of potential use, to refine those substances and to ensure that when they are manufactured for sale that the chemical composition remains consistent. The analytical methods used in these area can be very complex and may be specific to the process or area of research being conducted and may involve the use of bespoke analytical equipment.

FORENSIC ANALYSIS

In environmental management, water analysis is frequently deployed when pollution is suspected to identify the pollutant in order to take remedial action. The analysis can often enable the polluter to be identified. Such forensic work can examine the ratios of various components and can "type" samples of oils or other mixed organic contaminants to directly link the pollutant with the source. In drinking water supplies the cause of unacceptable quality can similarly be determined by carefully targeted chemical analysis of samples taken throughout the distribution system. In manufacturing, off-spec products may be directly tied back to unexpected changes in wet processing stages and analytical chemistry can identify which stages may be at fault and for what reason.

METHODOLOGY

To ensure consistency and repeatability, the methods use in the chemical analysis of water samples are often agreed and published at a national or state level. By convention these are often referred to as "Blue book"s

The methods defined in the relevant standards can be broadly classified as:

- **Conventional wet chemistry** including Winkler method method for dissolved oxygen , precipitation , filtration for solids, acidification, neutralisation titration etc. Colourimetric methods such as MBAS assay which indicates anionic surfactants in water and on site comparator methods to determine Chlorine and Chloramines. Nephelometers are used to measure solids concentrations as turbidity. These methods are generally robust and well tried and inexpensive, giving a reasonable degree of accuracy at modest sensitivity.
- **Electro chemistry** including pH, Conductivity and dissolved oxygen using oxygen electrode. These methods yield accurate and precise results using electronic equipment capable of feeding results directly into a laboratory data management system
- Spectrophotometry particularly for metallic elements in solution producing results with very high sensitivity but which may require some sample preparation prior to analysis and may also need specialised sampling methods to avoid sample deterioration in transit.
- Chromatography is used for many organic species which are volatile or which can yield a characteristic volatile component of after initial chemical processing.
- Ion chromatography is a sensitive and stable technique that can measure Li, NH_4 and many other low molecular weight ions using ion exchange technology.
- Gas chromatography can be used to determine methane, carbon dioxide, Cyanide oxygen and nitrogen and many other volatile components at reasonable sensitivities.
- Mass spectrometry is used where very high sensitivity is required and is sometimes used as a back-end process after Gas liquid chromatography for detecting trace organic chemicals.

Depending on the components, different methods are applied to determine the quantities or ratios of the components. While some methods can be performed with standard laboratory equipment, others require advanced devices, such as Inductively coupled plasma mass spectrometry (ICP-MS).

GAS CHROMATOGRAPHY

Gas chromatography–mass spectrometry (**GC-MS**) is an analytical method that combines the features of gas-chromatography and mass spectrometry to identify different substances within a test sample. Applications of GC-MS include drug detection, fire investigation, environmental analysis, explosives investigation, and identification of unknown samples, including that of material samples obtained from planet Mars during probe missions as early as the 1970s. GC-MS can also be used in airport security to detect substances in luggage or on human beings. Additionally, it can identify trace elements in materials that were previously thought to have disintegrated beyond identification. Like liquid chromatography–mass spectrometry, it allows analysis and detection even of tiny amounts of a substance.

GC-MS has been widely heralded as a "gold standard" for forensic substance identification because it is used to perform a 100% specific test, which positively identifies the presence of a particular substance. A nonspecific test merely indicates that any of several in a category of substances is present. Although a nonspecific test could statistically suggest the identity of the substance, this could lead to false positive identification.

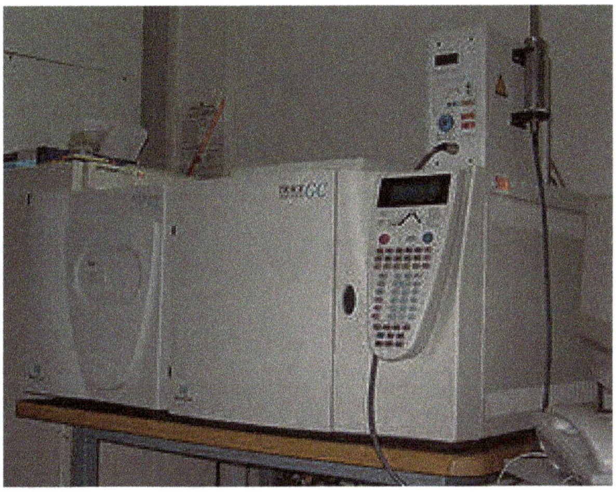

Example of a GC-MS instrument.

HISTORY

The use of a mass spectrometer as the detector in gas chromatography was developed during the 1950s after being originated by James and Martin in 1952. These comparatively sensitive devices were originally limited to laboratory settings.

The development of affordable and miniaturized computers has helped in the simplification of the use of this instrument, as well as allowed great improvements in the amount of time it takes to analyze a sample. In 1964, Electronic Associates, Inc. (EAI), a leading U.S. supplier of analog computers, began development of a computer controlled quadrupole mass spectrometer under the direction of Robert E. Finnigan. By 1966 Finnigan and collaborator Mike Uthe's EAI division had sold over 500 quadrupole residual gas-analyzer instruments. In 1967, Finnigan left EAI to form the Finnigan Instrument Corporation along with Roger Sant, T. Z. Chou, Michael Story, and William Fies. In early 1968, they delivered the first prototype quadrupole GC/MS instruments to Stanford and Purdue University. When Finnigan Instrument Corporation was acquired by Thermo Instrument Systems (later Thermo Fisher Scientific) in 1990, it was considered "the world's leading manufacturer of mass spectrometers".

In 1996 the top-of-the-line high-speed GC-MS units completed analysis of fire accelerants in less than 90 seconds, whereas first-generation GC-MS would have required at least 16 minutes. By the 2000s computerized GC/MS instruments using quadrupole technology had become both essential to chemical research and one of the foremost instruments used for organic analysis. Today computerized GC/MS instruments are widely used in environmental monitoring of water, air, and soil; in the regulation of agriculture and food safety; and in the discovery and production of medicine.

INSTRUMENTATION

The insides of the GC-MS, with the column of the gas chromatograph in the oven on the right.

The GC-MS is composed of two major building blocks: the gas chromatograph and the mass spectrometer. The gas chromatograph utilizes a capillary column which depends on the column's dimensions (length, diameter, film thickness) as well as the phase properties (e.g. 5% phenyl polysiloxane). The difference in the chemical properties between different molecules in a mixture and their relative affinity for the stationary phase of the column will promote separation of the molecules as the sample travels the length of the column. The molecules are retained by the column and then elute (come off) from the column at different times (called the retention time), and this allows the mass spectrometer downstream to capture, ionize, accelerate, deflect, and detect the ionized molecules separately. The mass spectrometer does this by breaking each molecule into ionized fragments and detecting these fragments using their mass-to-charge ratio.

These two components, used together, allow a much finer degree of substance identification than either unit used separately. It is not possible to make an accurate identification of a particular molecule by gas chromatography or mass spectrometry alone. The mass spectrometry process normally requires a very pure sample while gas chromatography using a traditional detector (e.g. Flame ionization detector) cannot differentiate between multiple molecules that happen to take the same amount of time to travel through the column (*i.e.* have the same retention time), which results in two or more molecules that co-elute. Sometimes two different molecules can also have a similar pattern of ionized fragments in a mass spectrometer (mass spectrum). Combining the two processes reduces the possibility of error, as it is extremely unlikely that two different molecules will behave in the same way in both a gas chromatograph and a mass spectrometer. Therefore, when an identifying mass spectrum appears at a characteristic retention time in a GC-MS analysis, it typically increases certainty that the analyte of interest is in the sample.

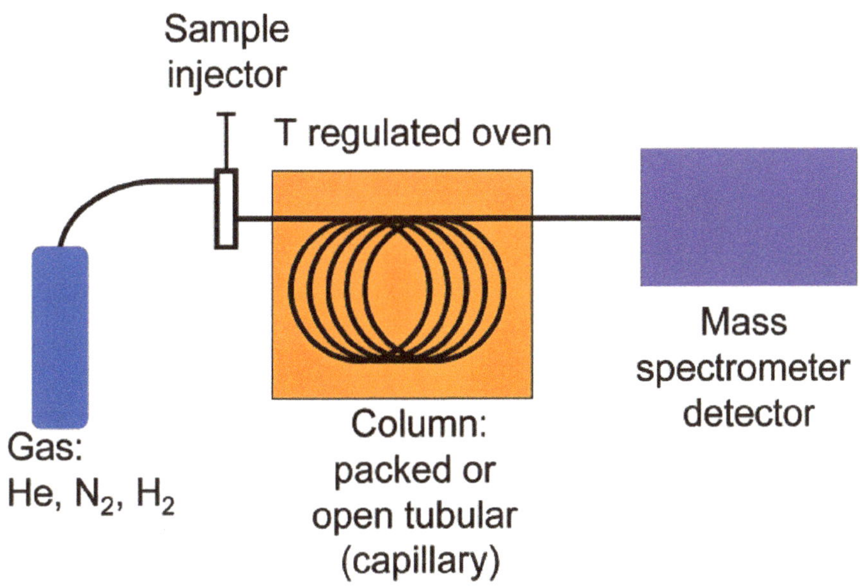

GC-MS schematic.

Purge and trap GC-MS

For the analysis of volatile compounds, a purge and trap (P&T) concentrator system may be used to introduce samples. The target analytes are extracted and mixed with water and introduced into an airtight chamber. An inert gas such as Nitrogen (N_2) is bubbled through the water; this is known as purging or sparging. The volatile compounds move into the headspace above the water and are drawn along a pressure gradient (caused by the introduction of the purge gas) out of the chamber. The volatile compounds are drawn along a heated line onto a 'trap'. The trap is a column of adsorbent material at ambient temperature that holds the compounds by returning them to the liquid phase. The trap is then heated and the sample compounds are introduced to the GC-MS column via a volatiles interface, which is a split inlet system. P&T GC-MS is particularly suited to volatile organic compounds (VOCs) and BTEX compounds (aromatic compounds associated with petroleum).

A faster alternative is the **"purge-closed loop"** system. In this system the inert gas is bubbled through the water until the concentrations of organic compounds in the vapor phase are at equilibrium with concentrations in the aqueous phase. The gas phase is then analysed directly.

Types of Mass Spectrometer Detectors

The most common type of mass spectrometer (MS) associated with a gas chromatograph (GC) is the quadrupole mass spectrometer, sometimes referred to by the Hewlett-Packard (now Agilent) trade name "Mass Selective Detector" (MSD). Another relatively common detector is the ion trap mass spectrometer. Additionally one may find a magnetic sector mass spectrometer, however these particular instruments are expensive and bulky and not typically found in high-throughput service laboratories. Other detectors may be encountered such as time of flight (TOF), tandem quadrupoles (MS-MS) (see below), or in the case of an ion trap MS^n where n indicates the number mass spectrometry stages.

GC-tandem MS

When a second phase of mass fragmentation is added, for example using a second quadrupole in a quadrupole instrument, it is called tandem MS (MS/MS). MS/MS can sometimes be used to quantitate low levels of target compounds in the presence of a high sample matrix background.

The first quadrupole (Q1) is connected with a collision cell (Q2) and another quadrupole (Q3). Both quadrupoles can be used in scanning or static mode, depending on the type of MS/MS analysis being performed. Types of analysis include product ion scan, precursor ion scan, selected reaction monitoring (SRM) (sometimes referred to as multiple reaction monitoring (MRM)) and neutral loss scan. For example: When Q1 is in static mode (looking at one mass only as in SIM), and Q3 is in scanning mode, one obtains a so-called product ion spectrum (also calleddaughter spectrum). From this spectrum, one can select a prominent product ion which can be the product ion for the chosen precursor ion. The pair is called atransition and forms the basis for SRM. SRM is highly specific and virtually eliminates matrix background.

Ionization

After the molecules travel the length of the column, pass through the transfer line and enter into the mass spectrometer they are ionized by various methods with typically only one method being used at any given time. Once the sample is fragmented it will then be detected, usually by an electron multiplier diode, which essentially turns the ionized mass fragment into an electrical signal that is then detected.

The ionization technique chosen is independent of using full scan or SIM.

Electron Ionization

By far the most common and perhaps standard form of ionization is electron ionization (EI). The molecules enter into the MS (the source is a quadrupole or the ion trap itself in an ion trap MS) where they are bombarded with free electrons emitted from a filament, not unlike the filament one would find in a standard light bulb. The electrons bombard the molecules, causing the molecule to fragment in a characteristic and reproducible way. This "hard ionization" technique results in the creation of more fragments of low mass to charge ratio (m/z) and few, if any, molecules approaching the molecular mass unit. Hard ionization is considered by mass spectrometrists as the employ of molecular electron bombardment, whereas "soft ionization" is charge by molecular collision with an introduced gas. The molecular fragmentation pattern is dependent upon the electron energy applied to the system, typically 70 eV (electron Volts). The use of 70 eV facilitates comparison of generated spectra with library spectra using manufacturer-supplied software or software developed by the National Institute of Standards (NIST-USA). Spectral library searches employ matching algorithms such as Probability Based Matching and dot-product matching that are used with methods of analysis written by many method standardization agencies. Sources of libraries include NIST, Wiley, the AAFS, and instrument manufacturers.

Cold Electron Ionization

The "hard ionization" process of electron ionization can be softened by the cooling of the molecules before their ionization, resulting in mass spectra that are richer in information. In this method named cold electron ionization (Cold-EI) the molecules exit the GC column, mixed with added helium make up gas and expand into vacuum through a specially designed supersonic nozzle, forming a supersonic molecular beam (SMB). Collisions with the make up gas at the expanding supersonic jet reduce the internal vibrational (and rotational) energy of the analyte molecules, hence reducing the degree of fragmentation caused by the electrons during the ionization process. Cold-EI mass spectra are characterized by an abundant molecular ion while the usual fragmentation pattern is retained, thus making Cold-EI mass spectra compatible with library search identification techniques. The enhanced molecular ions increase the identification probabilities of both known and unknown compounds, amplify isomer mass spectral effects and enable the use of isotope abundance analysis for the elucidation of elemental formulae.

Chemical Ionization

In chemical ionization a reagent gas, typically methane or ammonia is introduced into the mass spectrometer. Depending on the technique (positive CI or negative CI) chosen, this reagent gas will interact with the electrons and analyte and cause a 'soft' ionization of the molecule of interest. A softer ionization fragments the molecule to a lower degree than the hard ionization of EI. One of the main benefits of using chemical ionization is that a mass fragment closely corresponding to the molecular weight of the analyte of interest is produced.

In positive chemical ionization (PCI) the reagent gas interacts with the target molecule, most often with a proton exchange. This produces the species in relatively high amounts.

In negative chemical ionization (NCI) the reagent gas decreases the impact of the free electrons on the target analyte. This decreased energy typically leaves the fragment in great supply.

Analysis

A mass spectrometer is typically utilized in one of two ways: full scan or selected ion monitoring (SIM). The typical GC-MS instrument is capable of performing both functions either individually or concomitantly, depending on the setup of the particular instrument.

The primary goal of instrument analysis is to quantify an amount of substance. This is done by comparing the relative concentrations among the atomic masses in the generated spectrum. Two kinds of analysis are possible, comparative and original. Comparative analysis essentially compares the given spectrum to a spectrum library to see if its characteristics are present for some sample in the library. This is best performed by a computer because there are a myriad of visual distortions that can take place due to variations in scale. Computers can also simultaneously correlate more data (such as the retention times identified by GC), to more accurately relate certain data.

Another method of analysis measures the peaks in relation to one another. In this method, the tallest peak is assigned 100% of the value, and the other peaks being assigned proportionate values. All values above 3% are assigned. The total mass of the unknown compound is normally indicated by the parent peak. The value of this parent peak can be used to fit with a chemical formula containing the various elements which are believed to be in the compound. The isotope pattern in the spectrum, which is unique for elements that have many isotopes, can also be used to identify the various elements present. Once a chemical formula has been matched to the spectrum, the molecular structure and bonding can be identified, and must be consistent with the characteristics recorded by GC-MS. Typically, this identification done automatically by programs which come with the instrument, given a list of the elements which could be present in the sample.

A "full spectrum" analysis considers all the "peaks" within a spectrum. Conversely, selective ion monitoring (SIM) only monitors selected ions associated with a specific substance. This is done on the assumption that at a given retention time, a set of ions is characteristic of a certain compound. This is a fast and efficient analysis, especially if the analyst has previous information about a sample or is only looking for a few specific substances. When the amount of information collected about the ions in a given gas chromatographic peak decreases, the sensitivity of the analysis increases. So, SIM analysis allows for a smaller quantity of a compound to be detected and measured, but the degree of certainty about the identity of that compound is reduced.

Full scan MS

When collecting data in the full scan mode, a target range of mass fragments is determined and put into the instrument's method. An example of a typical broad range of mass fragments to monitor would be m/z 50 to m/z 400. The determination of what range to use is largely dictated by what one anticipates being in the sample while being cognizant of the solvent and other possible interferences. A MS should not be set to look for mass fragments too low or else one may detect air (found as *m/z* 28 due to nitrogen), carbon dioxide (*m/z* 44) or other possible interference. Additionally if one is to use a large scan range then sensitivity of the instrument is decreased due to performing fewer scans per second since each scan will have to detect a wide range of mass fragments.

Full scan is useful in determining unknown compounds in a sample. It provides more information than SIM when it comes to confirming or resolving compounds in a sample. During instrument method development it may be common to first analyze test solutions in full scan mode to determine the retention time and the mass fragment fingerprint before moving to a SIM instrument method.

Selected ion Monitoring

In selected ion monitoring (SIM) certain ion fragments are entered into the instrument method and only those mass fragments are detected by the mass spectrometer. The advantages of SIM are that the detection limit is lower since the instrument is only looking at a small number of fragments (e.g. three fragments) during each scan. More scans can take place each second. Since only a few mass fragments of interest are being monitored, matrix interferences are typically lower. To additionally confirm the likelihood of a potentially positive result, it is relatively important to be sure that the ion ratios of the various mass fragments are comparable to a known reference standard.

APPLICATIONS

Environmental Monitoring and Cleanup

GC-MS is becoming the tool of choice for tracking organic pollutants in the environment. The cost of GC-MS equipment has decreased significantly, and the reliability has increased at the same time, which has contributed to its increased adoption in environmental studies.

Criminal Forensics

GC-MS can analyze the particles from a human body in order to help link a criminal to a crime. The analysis of fire debris using GC-MS is well established, and there is even an established American Society for Testing and Materials (ASTM) standard for fire debris analysis. GCMS/MS is especially useful here as samples often contain very complex matrices and results, used in court, need to be highly accurate.

Law Enforcement

GC-MS is increasingly used for detection of illegal narcotics, and may eventually supplant drug-sniffing dogs. It is also commonly used in forensic toxicology to find drugs and/or poisons in biological specimens of suspects, victims, or the deceased.

Sports anti-Doping Analysis

GC-MS is the main tool used in sports anti-doping laboratories to test athletes' urine samples for prohibited performance-enhancing drugs, for example anabolic steroids.

Security

A post–September 11 development, explosive detection systems have become a part of all US airports. These systems run on a host of technologies, many of them based on GC-MS. There are only three manufacturers certified by the FAA to provide these systems, one of which is Thermo Detection (formerly Thermedics), which produces the EGIS, a GC-MS-based line of explosives detectors. The other two manufacturers are Barringer Technologies, now owned by Smith 's Detection Systems, and Ion Track Instruments, part of General Electric Infrastructure Security Systems.

Chemical Warfare Agent Detection

As part of the post-September 11 drive towards increased capability in homeland security and public health preparedness, traditional GC-MS units with transmission quadrupole mass spectrometers, as well as those with cylindrical ion trap (CIT-MS) and toroidal ion trap (T-ITMS) mass spectrometers have been modified for field portability and near real-time detection of chemical warfare agents (CWA) such as sarin, soman, and VX. These complex and large GC-MS systems have been modified and configured with resistively heated low thermal mass (LTM) gas chromatographs that reduce analysis time to less than ten percent of the time required in traditional laboratory systems. Additionally, the systems are smaller, and more mobile, including units that are mounted in mobile analytical laboratories (MAL), such as those used by the United States Marine Corps Chemical and Biological Incident Response Force MAL and other similar laboratories, and systems that are hand-carried by two-person teams or individuals, much ado to the smaller mass detectors. Depending on the system, the analytes can be introduced via liquid injection, desorbed from sorbent tubes through a thermal desorption process, or with solid-phase micro extraction (SPME).

Chemical Engineering

GC-MS is used for the analysis of unknown organic compound mixtures. One critical use of this technology is the use of GC-MS to determine the composition of bio-oils processed from raw biomass.

Food, beverage and perfume analysis

Foods and beverages contain numerous aromatic compounds, some naturally present in the raw materials and some forming during processing. GC-MS is extensively used for the analysis of these compounds which include esters, fatty acids, alcohols, aldehydes, terpenes etc. It is also used to detect and measure contaminants from spoilage or adulteration which may be harmful and which is often controlled by governmental agencies, for example pesticides.

Astrochemistry

Several GC-MS have left earth. Two were brought to Mars by the Viking program. Venera 11 and 12 and Pioneer Venus analysed the atmosphere of Venus with GC-MS. The Huygens probe of the Cassini-Huygens mission landed one GC-MS on Saturn's largest moon, Titan. The material in the comet 67P/Churyumov-Gerasimenko will be analysed by the Rosetta mission with a chiral GC-MS in 2014.

Medicine

Dozens of congenital metabolic diseases also known as Inborn error of metabolism are now detectable by newborn screening tests, especially the testing using gas chromatography–mass spectrometry. GC-MS can determine compounds in urine even in minor concentration. These compounds are normally not present but appear in individuals suffering with metabolic disorders. This is increasingly becoming a common way to diagnose IEM for earlier diagnosis and institution of treatment eventually leading to a better outcome. It is now possible to test a newborn for over 100 genetic metabolic disorders by a urine test at birth based on GC-MS.

In combination with isotopic labeling of metabolic compounds, the GC-MS is used for determining metabolic activity. Most applications are based on the use of ^{13}C as the labeling and the measurement of ^{13}C-^{12}C ratios with an **isotope ratio mass spectrometer (IRMS)**; an MS with a detector designed to measure a few select ions and return values as ratios.

LIQUID CHROMATOGRAPHY–MASS SPECTROMETRY

Liquid chromatography–mass spectrometry (**LC-MS**, or alternatively **HPLC-MS**) is an analytical chemistry technique that combines the physical separation capabilities of liquid chromatography (or HPLC) with the mass analysis capabilities of mass spectrometry (MS). LC-MS is a powerful technique that has very high sensitivity, making it useful in many applications. Its application is oriented towards the separation, general detection and potential identification of chemicals of particular masses in the presence of other chemicals (i.e., in complex mixtures), e.g., natural products from natural-products extracts, and pure substances from mixtures of chemical intermediates. Preparative LC-MS systems can be used for rapid mass-directed purification of specific substances from such mixtures that are important in basic research, and pharmaceutical, agrochemical, food, and other industries. Like gas chromatography–mass spectrometry, it allows analysis and detection even of tiny amounts of a substance.

Liquid Chromatography

Present day liquid chromatography generally utilizes very small particles packed and operating at relatively high pressure, and is referred to as high performance liquid chromatography (HPLC); modern LC-MS methods use HPLC instrumentation, essentially exclusively, for sample introduction. In HPLC, the sample is forced by a liquid at high pressure (the mobile phase) through a column that is packed with a stationary phase generally composed of irregularly or spherically shaped particles chosen or derivatized to accomplish particular types of separations. HPLC methods are historically divided into two different sub-classes based on stationary phases and the corresponding required polarity of the mobile phase. Use of octadecylsilyl (C18) and related organic-modified particles as stationary phase with pure or pH-adjusted water-organic mixtures such as water-acetonitrile and water-methanol are used in techniques termed reversed phase liquid chromatography (RP-LC). Use of materials such as silica gel as stationary phase with neat or mixed organic mixtures are used in techniques termed normal phase liquid chromatography (NP-LC). RP-LC is most often used as the means to introduce samples into the MS, in LC-MS instrumentation.

Flow Splitting

When standard bore (4.6 mm) columns are used the flow is often split ~10:1. This can be beneficial by allowing the use of other techniques in tandem such as MS and UV detection. However splitting the flow to UV will decrease the sensitivity of spectrophotometric detectors. The mass spectrometry on the other hand will give improved sensitivity at flow rates of 200 μL/min or less.

Mass Spectrometry

Mass spectrometry (MS) is an analytical technique that measures the mass-to-charge ratio of charged particles. It is used for determining masses of particles, for determining the elemental composition of a sample or molecule, and for elucidating the chemical structures of molecules, such as peptides and other chemical compounds. MS works by ionizing chemical compounds to generate charged molecules or molecule fragments and measuring their mass-to-charge ratios. In a typical MS procedure, a sample is loaded onto the MS instrument and undergoes vaporization. The components of the sample are ionized by one of a variety of methods (e.g., by impacting them with an electron beam, using uv lights as a photon beam, laser beam or corona discharge etc.), which results in the formation of charged particles (ions). The ions are separated according to their mass-to-charge ratio in an analyzer by electromagnetic fields. The ions are detected, usually by a quantitative method. The ion signal is processed into mass spectra.

Additionally, MS instruments consist of three modules. An ion source, which can convert gas phase sample molecules into ions (or, in the case of electrospray ionization, move ions that exist in solution into the gas phase). A mass analyzer, which sorts the ions by their masses by applying electromagnetic fields. A detector, which measures the value of an indicator quantity and thus provides data for calculating the abundances of each ion present

The technique has both qualitative and quantitative uses. These include identifying unknown compounds, determining the isotopic composition of elements in a molecule, and determining the structure of a compound by observing its fragmentation. Other uses include quantifying the amount of a compound in a sample or studying the fundamentals of gas phase ion chemistry (the chemistry of ions and neutrals in a vacuum). MS is now in very common use in analytical laboratories that study physical, chemical, or biological properties of a great variety of compounds.

Mass Analyzer

There are many different mass analyzers that can be used in LC/MS. Single quadrupole, triple quadrupole, ion trap, time of flight (TOF) and quadrupole-time of flight (Q-TOF).

Interface

Understandably the interface between a liquid phase technique which continuously flows liquid, and a gas phase technique carried out in a vacuum was difficult for a long time. The advent of electrospray ionization changed this. The interface is most often an electrospray ion source or variant such as a nanospray source; however atmospheric pressure chemical ionization interface is also used. Various deposition and drying techniques have also been used such as using moving belts; however the most common of these is off-line MALDI deposition. A new approach still under development called Direct-EI LC-MS interface, couples a nano HPLC system and an electron ionization equipped mass spectrometer.

Applications

Pharmacokinetics

LC-MS is very commonly used in pharmacokinetic studies of pharmaceuticals and is thus the most frequently used technique in the field of bioanalysis. These studies give information about how quickly a drug will be cleared from the hepatic blood flow, and organs of the body. MS is used for this due to high sensitivity and exceptional specificity compared to UV (as long as the analyte can be suitably ionised), and short analysis time.

The major advantage MS has is the use of tandem MS-MS. The detector may be programmed to select certain ions to fragment. The process is essentially a selection technique, but is in fact more complex. The measured quantity is the sum of molecule fragments chosen by the operator. As long as there are no interferences or ion suppression, the LC separation can be quite quick.

Proteomics/Metabolomics

LC-MS is also used in proteomics where again components of a complex mixture must be detected and identified in some manner. The bottom-up proteomics LC-MS approach to proteomics generally involves protease digestion and denaturation (usually trypsin as a protease, urea to denature tertiary structure and iodoacetamide to cap cysteine residues) followed by LC-MS with peptide mass fingerprinting or LC-MS/MS (tandem MS) to derive sequence of individual peptides. LC-MS/MS is most commonly used for proteomic analysis of complex samples where peptide masses may overlap even with a high-resolution mass spectrometer. Samples of complex biological fluids like human serum may be run in a modern LC-MS/MS system and result in over 1000 proteins being identified, provided that the sample was first separated on an SDS-PAGE gel or HPLC-SCX.

Profiling of secondary metabolites in plants or food like phenolics can be achieved with liquid chromatography–mass spectrometry.

Drug Development

LC-MS is frequently used in drug development at many different stages including peptide mapping, glycoprotein mapping, natural products dereplication, bioaffinity screening, *in vivo* drug screening, metabolic stability screening, metabolite identification, impurity identification, quantitative bioanalysis, and quality control.

PYROLYSIS–GAS CHROMATOGRAPHY–MASS SPECTROMETRY

Pyrolysis–gas chromatography–mass spectrometry is a method of chemical analysis in which the sample is heated to decomposition to produce smaller molecules that are separated by gas chromatography and detected using mass spectrometry.

How it works

Pyrolysis is the thermal decomposition of materials in an inert atmosphere or a vacuum. The sample is put into direct contact with a platinum wire, or placed in a quartz sample tube, and rapidly heated to 600–1000 °C. Depending on the application even higher temperatures are used. Three different heating techniques are used in actual pyrolyzers: Isothermal furnace, inductive heating (Curie Point filament), and resistive heating using platinum filaments. Large molecules cleave at their weakest points and produce smaller, more volatile fragments. These fragments can be separated by gas chromatography. Pyrolysis GC chromatograms are typically complex because a wide range of different decomposition products is formed. The data can either be used as fingerprint to prove material identity or the GC/MS data is used to identify individual fragments to obtain structural information. To increase the volatility of polar fragments, various methylating reagents can be added to a sample before pyrolysis.

Besides the usage of dedicated pyrolyzers, pyrolysis GC of solid and liquid samples can be performed directly inside Programmable Temperature Vaporizer (PTV) injectors that provide quick heating (up to 60 °C/s) and high maximum temperatures of 600-650 °C. This is sufficient for many pyrolysis applications. The main advantage is that no dedicated instrument has to be purchased and pyrolysis can be performed as part of routine GC analysis. In this case quartz GC inlet liners can be used. Quantitative data can be acquired, and good results of derivatization inside the PTV injector are published as well.

Applications

Pyrolysis gas chromatography is useful for the identification of involatile compounds and synthetic polymeric media, such as acrylics or alkyds, and synthetic varnishes. The way in which the polymer fragments before it is separated in the GC can help in identifying the way that the sample polymer assembled itself (step growth v. chain growth, for example). The method of pyrolysis gas chromatography can also be used for environmental samples, including fossils. Trace evidence scientists in forensic laboratories use Pyroylsis GC to compare and identify paint or finish found at crime scenes or victims.

AFFINITY CHROMATOGRAPHY

Affinity chromatography is a method of separating biochemical mixtures based on a highly specific interaction such as that between antigen and antibody, enzyme and substrate, or receptor and ligand.

Uses

Affinity chromatography can be used to:

- Purify and concentrate a substance from a mixture into a buffering solution
- Reduce the amount of a substance in a mixture
- Discern what biological compounds bind to a particular substance
- Purify and concentrate an enzyme solution.

Principle

The stationary phase is typically a gel matrix, often of agarose; a linear sugar molecule derived from algae. Usually the starting point is an undefined heterogeneous group of molecules in solution, such as a cell lysate, growth medium or blood serum. The molecule of interest will have a well known and defined property, and can be exploited during the affinity purification process. The process itself can be thought of as an entrapment, with the target molecule becoming trapped on a solid or stationary phase or medium. The other molecules in the mobile phase will not become trapped as they do not possess this property. The stationary phase can then be removed from the mixture, washed and the target molecule released from the entrapment in a process known as elution. Possibly the most common use of affinity chromatography is for the purification of recombinant proteins.

Batch and Column Setups

Binding to the solid phase may be achieved by column chromatography whereby the solid medium is packed onto a column, the initial mixture run through the column to allow setting, a wash buffer run through the column and the elution buffer subsequently applied to the column and collected. These steps are usually done at ambient pressure. Alternatively, binding may be achieved using a batch treatment, for example, by adding the initial mixture to the solid phase in a vessel, mixing, separating the solid phase, removing the liquid phase, washing, re-centrifuging, adding the elution buffer, re-centrifuging and removing the elute.

Sometimes a hybrid method is employed such that the binding is done by the batch method, but the solid phase with the target molecule bound is packed onto a column and washing and elution are done on the column.

A third method, expanded bed absorption, which combines the advantages of the two methods mentioned above, has also been developed. The solid phase particles are placed in a column where liquid phase is pumped in from the bottom and exits at the top. The gravity of the particles ensure that the solid phase does not exit the column with the liquid phase.

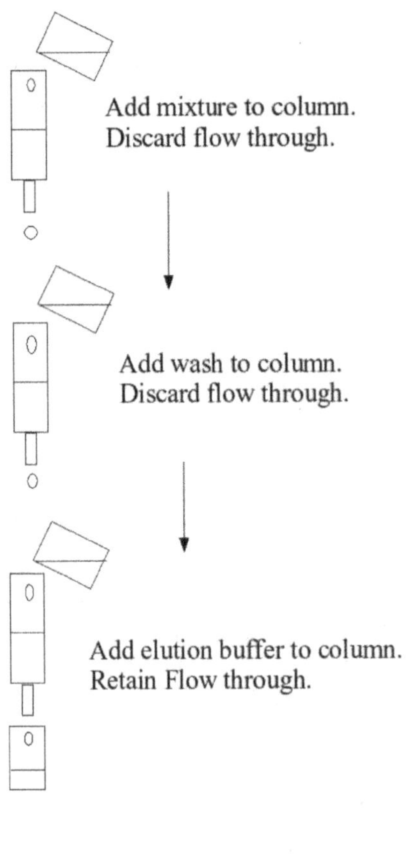

Column chromatography.

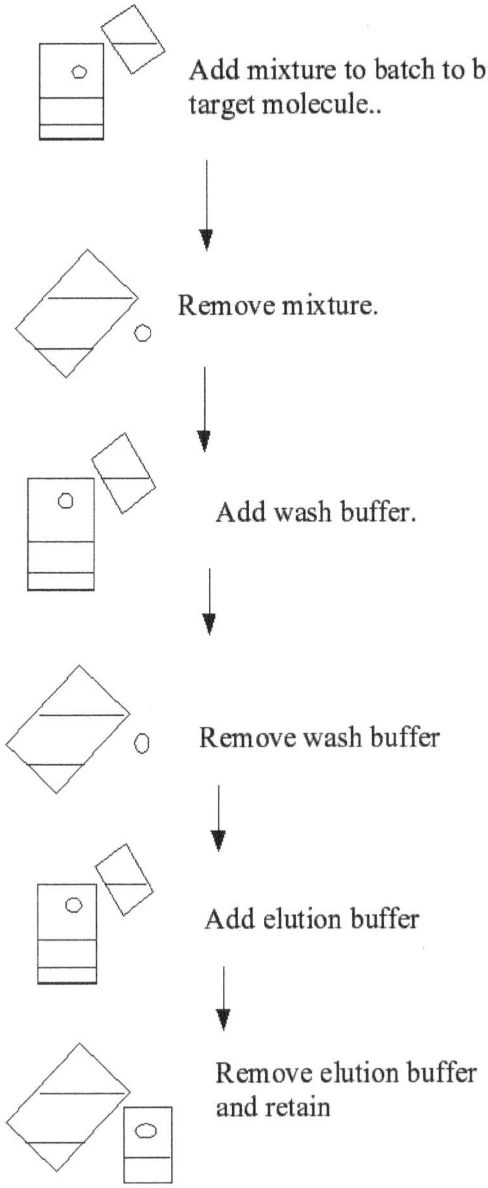

Batch chromatography.

Affinity columns can be eluted by changing salt concentrations, pH, pI, charge and ionic strength directly or through a gradient to resolve the particles of interest.

More recently, setups employing more than one column in series have been developed. The advantage compared to single column setups is that the resin material can be fully loaded, since non-binding product is directly passed on to a consecutive column with fresh column material. The resin costs per amount of produced product can thus be drastically be reduced. Since one column can always be eluted and regenerated while the other column is loaded, already two columns are sufficient to make full use of the advantages. Additional columns can give additional flexibility for elution and regeneration times, at the cost of additional equipment and resin costs.

Specific Uses

Affinity chromatography can be used in a number of applications, including nucleic acid purification, protein purification from cell free extracts, and purification from blood.

Various Affinity Media

Many different affinity media exist for a variety of possible uses. Briefly, they are (generalized):

- Activated/Functionalized – Works as a functional spacer, support matrix, and eliminates handling of toxic reagents.
- Amino Acid – Used with a variety of serum proteins, proteins, peptides, and enzymes, as well as rRNA and dsDNA.
- Avidin Biotin – Used in the purification process of biotin/avidin and their derivatives.
- Carbohydrate Bonding – Most often used with glycoproteins or any other carbohydrate-containing substance.
- Carbohydrate – Used with lectins, glycoproteins, or any other carbohydrate metabolite protein.
- Dye Ligand – This media is nonspecific, but mimics biological substrates and proteins.
- Glutathione – Useful for separation of GST tagged recombinant proteins.
- Heparin – This media is a generalized affinity ligand, and it is most useful for separation of plasma coagulation proteins, along with nucleic acid enzymes and lipases.
- Hydrophobic Interaction – Most commonly used to target free carboxyl groups and proteins.
- Immunoaffinity – Detailed below, this method utilizes antigens' and antibodies' high specificity to separate.
- Immobilized Metal Affinity Chromatography – Detailed further below, this method uses interactions between metal ions and proteins (usually specially tagged) to separate.
- Nucleotide/Coenzyme – Works to separate dehydrogenases, kinases, and transaminases.
- Nucleic Acid – Functions to trap mRNA, DNA, rRNA, and other nucleic acids/oligonucleotides.

- Protein A/G – This method is used to purify immunoglobulins.
- Speciality – Designed for a specific class or type of protein/coenzyme, this type of media will only work to separate a specific protein or coenzyme.

Immunoaffinity

Another use for the procedure is the affinity purification of antibodies from blood serum. If serum is known to contain antibodies against a specific antigen (for example if the serum comes from an organism immunized against the antigen concerned) then it can be used for the affinity purification of that antigen. This is also known as Immunoaffinity Chromatography. For example, if an organism is immunised against a GST-fusion protein it will produce antibodies against the fusion-protein, and possibly antibodies against the GST tag as well. The protein can then be covalently coupled to a solid support such as agarose and used as an affinity ligand in purifications of antibody from immune serum.

For thoroughness the GST protein and the GST-fusion protein can each be coupled separately. The serum is initially allowed to bind to the GST affinity matrix. This will remove antibodies against the GST part of the fusion protein. The serum is then separated from the solid support and allowed to bind to the GST-fusion protein matrix. This allows any antibodies that recognize the antigen to be captured on the solid support. Elution of the antibodies of interest is most often achieved using a low pH buffer such as glycine pH 2.8. The eluate is collected into a neutral tris or phosphate buffer, to neutralize the low pH elution buffer and halt any degradation of the antibody's activity. This is a nice example as affinity purification is used to purify the initial GST-fusion protein, to remove the undesirable anti-GST antibodies from the serum and to purify the target antibody.

A simplified strategy is often employed to purify antibodies generated against peptide antigens. When the peptide antigens are produced synthetically, a terminal cysteine residue is added at either the N- or C-terminus of the peptide. This cysteine residue contains a sulfhydryl functional group which allows the peptide to be easily conjugated to a carrier protein (e.g. Keyhole limpet hemocyanin (KLH)). The same cysteine-containing peptide is also immobilized onto an agarose resin through the cysteine residue and is then used to purify the antibody.

Most monoclonal antibodies have been purified using affinity chromatography based on immunoglobulin-specific Protein A or Protein G, derived from bacteria.

Immobilized metal ion Affinity Chromatography

Immobilized metal ion affinity chromatography (IMAC) is based on the specific coordinate covalent bond of amino acids, particularly histidine, to metals. This technique works by allowing proteins with an affinity for metal ions to be retained in a column containing immobilized metal ions, such as cobalt, nickel, copper for the purification of histidine containing proteins or peptides, iron, zinc or gallium for the purification of phosphorylated proteins or peptides. Many naturally occurring proteins do not have an affinity for metal ions, therefore recombinant DNA technology can be used to introduce such a protein tag into the relevant gene. Methods used to elute the protein of interest include changing the pH, or adding a competitive molecule, such as imidazole.

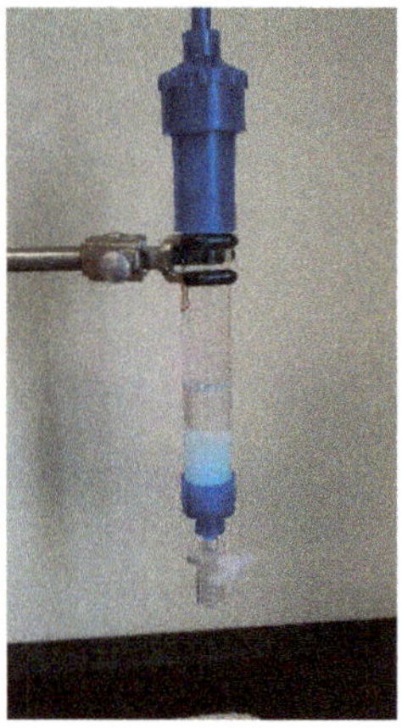

A chromatography column containing nickel-agarose beads used for purification of proteins with histidine tags

Recombinant Proteins

Possibly the most common use of affinity chromatography is for the purification of recombinant proteins. Proteins with a known affinity are protein tagged in order to aid their purification. The protein may have been genetically modified so as to allow it to be selected for affinity binding; this is known as a fusion protein. Tags include glutathione-S-transferase (GST), hexahistidine (His), and maltose binding protein (MBP). Histidine tags have an affinity for nickel or cobalt ions which have been immobilized by forming coordinate covalent bonds with a chelator incorporated in the stationary phase. For elution, an excess amount of a compound able to act as a metal ion ligand, such as imidazole, is used. GST has an affinity for glutathione which is commercially available immobilized as glutathione agarose. During elution, excess glutathione is used to displace the tagged protein.

Lectins

Lectin affinity chromatography is a form of affinity chromatography where lectins are used to separate components within the sample. Lectins, such as Concanavalin A are proteins which can bind specific carbohydrate (sugar) molecules. The most common application is to separate glycoproteins from non-glycosylated proteins, or one glycoform from another glycoform.

ANALYTICAL QUALITY CONTROL

Analytical quality control, commonly shortened to **AQC** refers to all those processes and procedures designed to ensure that the results of laboratory analysis are consistent, comparable, accurate and within specified limits of precision. Constituents submitted to the analytical laboratory must be accurately described to avoid faulty interpretations, approximations, or incorrect results. The qualitative and quantitative data generated from the laboratory can then be used for decision making. In the chemical sense, quantitative analysis refers to the measurement of the amount or concentration of an element or chemical compound in a matrix that differs from the element or compound. Fields such as industry, medicine, and law enforcement can make use of AQC.

In the Laboratory

AQC processes are of particular importance in laboratories analysing environmental samples where the concentration of chemical species present may be extremely low and close to the detection limit of the analytical method. In well managed laboratories, AQC processes are built into the routine operations of the laboratory often by the random introduction of known standards into the sample stream or by the use of spiked samples.

Quality control begins with sample collection and ends with the reporting of data. AQC is achieved through laboratory control of analytical performance. Initial control of the complete system can be achieved through specification of laboratory services, instrumentation, glassware, reagents, solvents, and gases. However, evaluation of daily performance must be documented to ensure continual production of valid data. A check should first be done to ensure that the data should be seen is precise and accurate. Next, systematic daily checks such as analysing blanks, calibration standards, quality control check samples, and references must be performed to establish the reproducibility of the data. The checks help certify that the methodology is measuring what is in the sample.

The quality of individual AQC efforts can be variable depending on the training, professional pride, and importance of a particular project to a particular analyst. The burden of an individual analyst originating AQC efforts can be lessened through the implementation of quality assurance programs. Through the implementation of established and routine quality assurance programs, two primary functions are fulfilled: the determination of quality, and the control of quality. By monitoring the accuracy and precision of results, the quality assurance program should increase confidence in the reliability of the reported analytical results, thereby achieving adequate AQC.

Pharmaceutical Industry

Validation of analytical procedures is imperative in demonstrating that a drug substance is suitable for a particular purpose. Common validation characteristics include: accuracy, precision (repeatability and intermediate precision), specificity, detection limit, quantitation limit, linearity, range, and robustness. In cases such as changes in synthesis of the drug substance, changes in composition of the finished product, and changes in the analytical procedure, revalidation is necessary to ensure quality control.

All analytical procedures should be validated. Identification tests are conducted to ensure the identity of an analyte in a sample through comparison of the sample to a reference standard through methods such as spectrum, chromatographic behavior, and chemical reactivity.Impurity testing can either be a quantitative test or a limit test. Both tests should accurately measure the purity of the sample. Quantitative tests of either the active moiety or other components of a sample can be conducted through assay procedures.

Other analytical procedures such as dissolution testing or particle size determination may also need to be validated and are equally important.

Statistics

Because of the complex inter-relationship between analytical method, sample concentration, limits of detection and method precision, the management of AQC is undertaken using a statistical approach to determine whether the results obtained lie within an acceptable statistical envelope.

Inter-laboratory Calibration

In circumstances where more than one laboratory is analysing samples and feeding data into a large programme of work such as the Harmonised monitoring scheme in the UK, AQC can also be applied to validate one laboratory against another. In such cases the work may be referred to as **inter-laboratory calibration**.

REFERENCES

1. James, A. T.; Martin, A. J. (1952). "Gas-liquid partition chromatography; the separation and micro-estimation of volatile fatty acids from formic acid to dodecanoic acid". The Biochemical Journal. 50 *(5): 679–90.* doi:10.1042/bj0500679. PMC 1197726. PMID 14934673.
2. Brock, David C. (2011).A Measure of Success. Chemical Heritage Magazine. 29 *(1). Retrieved 22 April 2014.*
3. Brock, David C. (2008).Detecting Success. Chemical Heritage Magazine. 26 *(2): 31.*
4. Thermo Instrument Systems Inc. History. International Directory of Company Histories (Volume 11 ed.). St. James Press. 1995. pp. 513–514. Retrieved 23 January 2015.
5. JeromeJeyakumar, J.; et. al. (April 2013).A Study of Phytochemical Constituents in Caralluma Umbellata By Gc – Ms Anaylsis [sic] (PDF). International Journal of Pharmaceutical Science Invention: 37–41. Retrieved 23 January 2015.
6. "Optimizing the Analysis of Volatile Organic Compounds – Technical Guide" Restek Corporation, Lit. Cat. 59887A
7. Wang, T.; Lenahan, R. (1984).Determination of volatile halocarbons in water by purge-closed loop gas chromatography. Bulletin of Environmental Contamination and Toxicology. 32 *(1): 429–438.* doi:10.1007/BF01607519. ISSN 0007-4861.
8. McLafferty, F. W.; Hertel, R. H.; Villwock, R. D. (1974).Probability based matching of mass spectra. Rapid identification of specific compounds in mixtures. Organic Mass Spectrometry. 9 *(7): 690–702. doi:10.1002/oms.1210090710.*
9. Stein, SE; Scott DR (1994).Optimization and testing of mass spectral library search algorithms for compound identification. J Am Soc Mass Spectrom. 5 *(9):*

859–866. doi:10.1016/1044-0305(94)87009-8. PMID 24222034.

10. Standard Reference Data. nist.gov
11. Wiley's Scientific, Technical, and Medical Databases: Home. wiley.com
12. Mass Spectrometry Database Committee. ualberta.ca
13. Amirav A, Gordin A, Poliak M, Fialkov AB (2008).Gas chromatography-mass spectrometry with supersonic molecular beams. J Mass Spectrom. 43 *(2): 141–63. doi:10.1002/jms.1380. PMID 18225851.*
14. SMB-MS (Supersonic GC-MS). tau.ac.il
15. Alon, T.; Amirav, A. (2006).Isotope Abundance Analysis Method and Software for Improved Sample Identification with the Supersonic GC-MS. Rapid Communications in Mass Spectrometry. 20 *(17): 2579–2588. doi:10.1002/rcm.2637. PMID 16897787.*
16. Tsivou, M.; Kioukia-Fougia, N.; Lyris, E.; Aggelis, Y.; Fragkaki, A.; Kiousi, X.; Simitsek, P.; Dimopoulou, H.; Leontiou, I. -P.; Stamou, M.; Spyridaki, M. -H.; Georgakopoulos, C. (2006).An overview of the doping control analysis during the Olympic Games of 2004 in Athens, Greece. Analytica Chimica Acta. 555: 1–13. doi:10.1016/j.aca.2005.08.068.
17. Smith, P. A.; Lepage, C. J.; Lukacs, M.; Martin, N.; Shufutinsky, A.; Savage, P. B. (2010).Field-portable gas chromatography with transmission quadrupole and cylindrical ion trap mass spectrometric detection: Chromatographic retention index data and ion/molecule interactions for chemical warfare agent identification. International Journal of Mass Spectrometry. 295 *(3): 113–118. doi:10.1016/j.ijms.2010.03.001.*
18. Sloan, K. M.; Mustacich, R. V.; Eckenrode, B. A. (2001).Development and evaluation of a low thermal mass gas chromatograph for rapid forensic GC-MS analyses. Field Analytical Chemistry & Technology. 5 *(6): 288–301. doi:10.1002/fact.10011.*
19. Patterson, G. E.; Guymon, A. J.; Riter, L. S.; Everly, M.; Griep-Raming, J.; Laughlin, B. C.; Ouyang, Z.; Cooks, R. G. (2002).Miniature Cylindrical Ion Trap Mass Spectrometer. Analytical Chemistry. 74 *(24): 6145–6153.* doi:10.1021/ac020494d. PMID 12510732.
20. Tekin, Kubilay; Karagöz, Selhan; Bektaş, Sema (2014-12-01).A review of hydrothermal biomass processing. Renewable and Sustainable Energy Reviews. 40: 673–687. doi:10.1016/j.rser.2014.07.216.
21. SEARCHING FOR LIFE ON MARS: The Development of the Viking GCMS. NASA
22. Krasnopolsky, V. A.; Parshev, V. A. (1981).Chemical composition of the atmosphere of Venus. Nature. 292 *(5824): 610–613.* doi:10.1038/292610a0.
23. Niemann, H. B.; Atreya, S. K.; Bauer, S. J.; Carignan, G. R.; Demick, J. E.; Frost, R. L.; Gautier, D.; Haberman, J. A.; Harpold, D. N.; Hunten, D. M.; Israel, G.; Lunine, J. I.; Kasprzak, W. T.; Owen, T. C.; Paulkovich, M.; Raulin, F.; Raaen, E.; Way, S. H. (2005).The abundances of constituents of Titan's

atmosphere from the GCMS instrument on the Huygens probe. Nature. 438 *(7069): 779–784.* doi:10.1038/nature04122. PMID 16319830.

24. Goesmann, F.; Rosenbauer, H.; Roll, R.; Böhnhardt, H. (2005).COSAC Onboard Rosetta: A Bioastronomy Experiment for the Short-Period Comet 67P/Churyumov-Gerasimenko. Astrobiology. 5 *(5): 622–631. Bibcode*:2005AsBio...5..622G. doi:10.1089/ast.2005.5.622. PMID 16225435.

CHAPTER 3

Soil-Forming Processes and Properties

Chapter Contents

INTRODUCTION

Soil is a mixture of minerals, organic matter, gases, liquids, and countless organisms that together support life on Earth. Soil is a natural body called the pedosphere which has four important functions: it is a medium for plant growth; it is a means of water storage, supply and purification; it is a modifier of Earth's atmosphere; it is a habitat for organisms; all of which, in turn, modify the soil.

Soil is called the Skin of the Earth and interfaces with the lithosphere, the hydrosphere, the atmosphere, and the biosphere. The term pedolith, used commonly to refer to the soil, literally translates ground stone. Soil consists of a solid phase of minerals (the soil matrix) and organic matter, as well as a porous phase that holds gases (the soil atmosphere) and water (the soil solution). Accordingly, soils are often treated as a three-state system of solids, liquids, and gases.

Soil is a product of the influence of climate, relief (elevation, orientation, and slope of terrain), organisms, and its parent materials (original minerals) interacting over time. Soil continually undergoes development by way of numerous physical, chemical and biological processes, which include weathering with associated erosion. Given its complexity and strong internal connectedness soil has been considered as an ecosystem by soil ecologists.

Most soils have a dry bulk density (density of soil taking into account voids when dry) between 1.1 and 1.6 g/cm3, while the soil particle density is much higher, in the range of 2.6 to 2.7 g/cm3. Little of the soil of planet Earth is older than the Pleistocene and none is older than the Cenozoic, although fossilized soils are preserved from as far back as the Archean.

Soil science has two basic branches of study: edaphology and pedology. Edaphology is concerned with the influence of soils on living things. Pedology is focused on the formation, description (morphology), and classification of soils in their natural environment. In engineering terms, soil is referred to as regolith, or loose rock material that lies above the solid geology. Soil is commonly referred to as earth or dirt; technically, the term dirt should be restricted to displaced soil.

As soil resources serve as a basis for food security, the international community advocates its sustainable and responsible use through different types of soil governance.

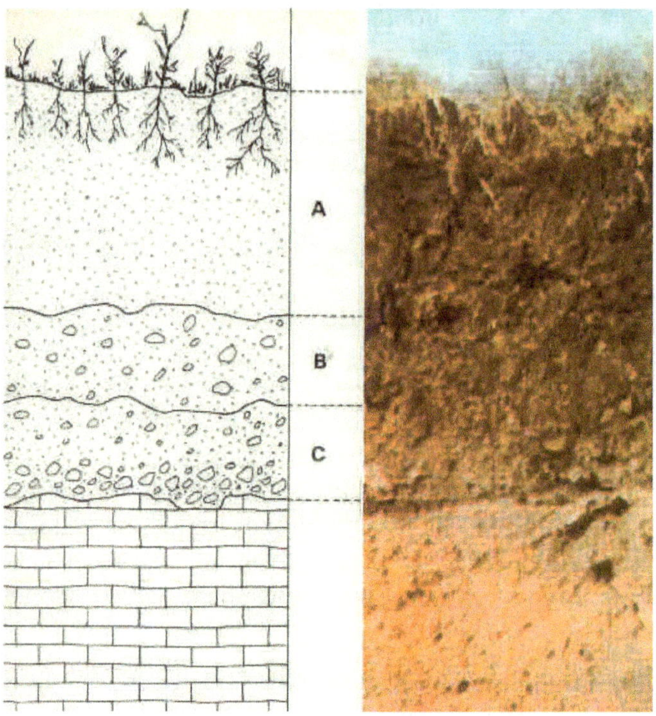

A, B, and C represent the soil profile, a notation firstly coined by Vasily Dokuchaev, the father of pedology; A is the topsoil; B is a regolith; C is a saprolite, a less-weathered regolith; the bottom-most layer represents the bedrock.

Surface-water-gley developed in glacial till, northern Ireland.

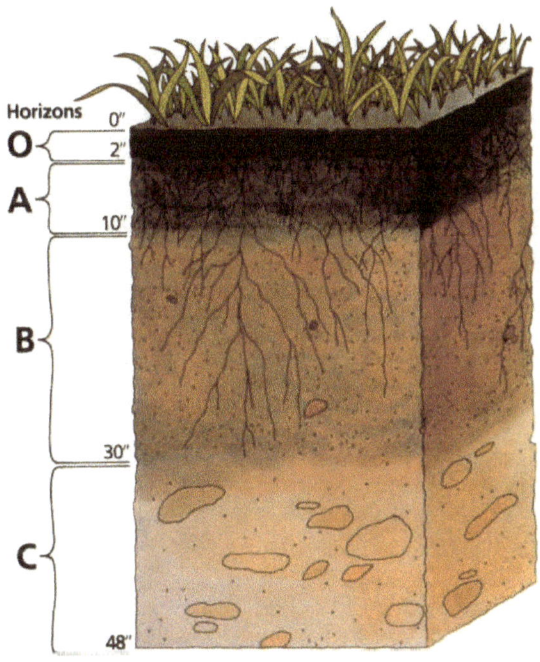

Soil Profile: Darkened topsoil and reddish subsoil layers are typical in some regions.

Soil is a major component of the Earth's ecosystem. The world's ecosystems are impacted in far-reaching ways by the processes carried out in the soil, from ozone depletion and global warming, to rainforest destruction and water pollution. Following the atmosphere, the soil is the next largest carbon reservoir on Earth, and it is potentially one of the most reactive to human disturbance and climate change. As the planet warms, soils will add carbon dioxide to the atmosphere due to its increased biological activity at higher temperatures. Thus, soil carbon losses likely have a large positive feedback response (amplification) to global warming, although positive feedback has been questioned on the base of more recent knowledge on soil carbon turnover.

Soil acts as an engineering medium, a habitat for soil organisms, a recycling system for nutrients and organic wastes, a regulator of water quality, a modifier of atmospheric composition, and a medium for plant growth, in other terms one of the world's best providers of ecosystem services. Since soil has a tremendous range of available niches and habitats, it contains most of the Earth's genetic diversity. A gram of soil can contain billions of organisms, belonging to thousands of species, mostly microbial and in the main still unexplored. Soil has a mean prokaryotic density of roughly 10^8organisms per gram, whereas the ocean has no more than 10^7 procaryotic organisms per milliliter (gram) of seawater.The carbon content of the soil is eventually returned to the atmosphere through the process of respiration carried out by heterotrophic organisms that feed upon the carbonaceous material in the soil, but a substantial part is retained in the soil in the form of humus, the ratio respired to fixed carbon decreasing with cultivation. Since plant roots need oxygen, ventilation is an important characteristic of soil. This ventilation can be accomplished via networks of interconnected soil pores, which also absorb and hold rainwater making it readily available for plant uptake. Since plants require a nearly continuous supply of water, but most regions receive sporadic rainfall, the water-holding capacity of soils is vital for plant survival.

Soils can effectively remove impurities, kill disease agents, and degrade contaminants, this latter property being called natural attenuation. Typically, soils maintain a net absorption of oxygen and methane, and undergo a net release of carbon dioxide and nitrous oxide. Soils offer plants physical support, air, water, temperature moderation, nutrients, and protection from toxins. Soils provide readily available nutrients to plants and animals by converting dead organic matter into various nutrient forms.

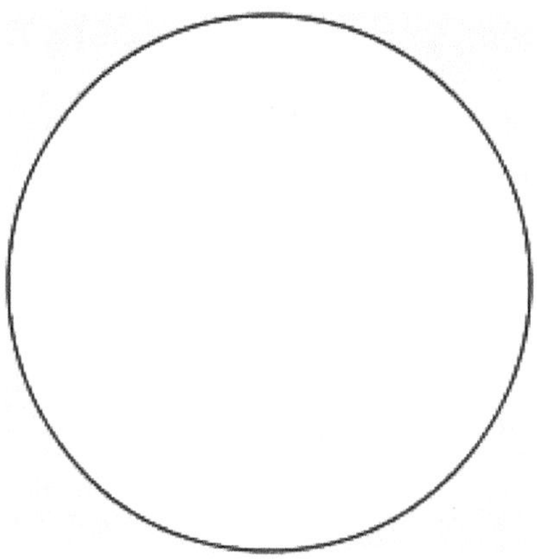

Components of a loam soil by percent volume

- Water (25%)
- Gases (25%)

- Sand (18%)
- Silt (18%)
- Clay (9%)
- Organic matter (5%)

Soils supply plants with nutrients that are held in place by their clay and humus content. For optimum plant growth, the soil components by volume should be roughly 50% solids (45% mineral and 5% organic matter), and 50% voids of which half is occupied by water and half by gas. The percent soil mineral and organic content is typically treated as a constant, while the percent soil water and gas content is considered highly variable whereby a rise in one is simultaneously balanced by a reduction in the other. The pore space allows for the infiltration and movement of air and water, both of which are critical for life in soil. Compaction, a common problem with soils, reduces this space, preventing air and water from reaching the plant roots and soil organisms.

Given sufficient time, an undifferentiated soil will evolve a soil profile which consists of two or more layers, referred to as soil horizons, that differ in one or more properties such as in their texture, structure, density, porosity, consistency, temperature, color, and reactivity. The horizons differ greatly in thickness and generally lack sharp boundaries. Soil profile development is dependent on the processes that form soils from their parent materials, the type of parent material, and the factors that control soil formation. The biological influences on soil properties are strongest near the surface, while the geochemical influences on soil properties increase with depth. Mature soil profiles in temperate climate regions typically include three basic master horizons: A, B and C. The solum normally includes the A and B horizons. The living component of the soil is largely confined to the solum. In the more hot, humid, climate of the tropics, a soil may have only a single horizon, when all the rock material has been converted to soil (residual soil).

The soil texture is determined by the relative proportions of sand, silt, and clay in the soil. The addition of organic matter, water, gases and time-related biotic and abiotic processes causes that texture to develop into a larger soil structure called an aggregate or a ped. At that point a soil can be said to be developed, and can be described further in terms of color, porosity, consistency, reaction (acidity), etc.

Of all the factors influencing the evolution of soil, water is the most powerful due to its involvement in the solution, erosion, transportation, and deposition of the materials of which a soil is composed. The mixture of water and the dissolved or suspended materials that occupy the soil pore space is called the soil solution. Since soil water is never pure water, but contains hundreds of dissolved organic and mineral substances, it may be more accurately called the soil solution. Water is central to the solution, precipitation and leaching of minerals from the soil profile. Finally, water affects the type of vegetation that grows in a soil, which in turn affects the development of the soil.

The most influential factor in stabilizing soil fertility are the soil colloids, clay and humus. Soil colloids behave as repositories of nutrients and moisture and so act to buffer the variations of soil solution ions and moisture. The contribution of soil colloids to soil nutrition are out of proportion to their part of the soil. Colloids act to store nutrients that might otherwise be leached from the soil or to release those ions in response to changes of soil pH, and so, make them available to plants.

The greatest influence on plant nutrient availability is soil pH, which is a measure of the hydrogen ion (acid-forming) soil reactivity, and is in turn a function of the soil materials, precipitation level, and plant root behavior. Soil pH strongly affects the availability of nutrients.

Most nutrients, with the exception of nitrogen, originate from minerals. Some nitrogen originates from rain (as dilute nitric acid), but most of the nitrogen is available in soils as a result of nitrogen fixation by bacteria. The action of microbes on organic matter and minerals may be to free nutrients for use, sequester them, or cause their loss from the soil by their volatilisation to gases or their leaching from the soil. The nutrients may be stored on soil colloids, or within live or dead organic matter, but they may not be accessible to plants due to extremes of pH.

The organic material of the soil has a powerful effect on its development, fertility, and available moisture. Following water and soil colloids, organic material is next in importance to a soil's formation and fertility.

History of the Study of Soil

Studies Concerning Soil Fertility

The history of the study of soil is intimately tied to our urgent need to provide food for ourselves and forage for our animals. Throughout history, civilizations have prospered or declined as a function of the availability and productivity of their soils.

The Greek historian Xenophon (450–355 B.C.) is credited with being the first to expound upon the merits of green-manuring crops: "But then whatever weeds are upon the ground, being turned into earth, enrich the soil as much as dung."

Columella's "Husbandry," circa 60 A.D., advocated the use of lime and that clover and alfalfa (green manure) should be turned under, and was used by 15 generations (450 years) under the Roman Empire until its collapse. From the fall of Rome to the French Revolution, knowledge of soil and agriculture was passed on from parent to child and as a result, crop yields were low. During the European Dark Ages, Yahya Ibn al-'Awwam's handbook, with its emphasis on irrigation, guided the people of North Africa, Spain and the Middle East; a translation of this work was finally carried to the southwest of the United States.

Experiments into what made plants grow first led to the idea that the ash left behind when plant matter was burned was the essential element but overlooked the role of nitrogen, which is not left on the ground after combustion. In about 1635, the Flemish chemist Jan Baptist van Helmont thought he had proved water to be the essential element from his famous five years' experiment with a willow tree grown with only the addition of rainwater. His conclusion came from the fact that the increase in the plant's weight had apparently been produced only by the addition of water, with no reduction in the soil's weight. John Woodward (d. 1728) experimented with various types of water ranging from clean to muddy and found muddy water the best, and so he concluded that earthy matter was the essential element. Others concluded it was humus in the soil that passed some essence to the growing plant. Still others held that the vital growth principal was something passed from dead plants or animals to the new plants. At the start of the 18th century, Jethro Tull demonstrated that it was beneficial to cultivate (stir) the soil, but his opinion that the stirring made the fine parts of soil available for plant absorption was erroneous.

As chemistry developed, it was applied to the investigation of soil fertility. The French chemist Antoine Lavoisier showed in about 1778 that plants and animals must "combust" oxygen internally to live and was able to deduce that most of the 165-pound weight of van Helmont's willow tree derived from air. It was the French agriculturalist Jean-Baptiste Boussingault who by means of experimentation obtained evidence showing that the main sources of carbon, hydrogen and oxygen for plants were the air and water. Justus von Liebig in his book *Organic Chemistry in its Applications to Agriculture and Physiology* (published 1840), asserted that the chemicals in plants must have come from the soil and air and that to maintain soil fertility, the used minerals must be replaced. Liebig nevertheless believed the nitrogen was supplied from the air. The enrichment of soil with guano by the Incas was rediscovered in 1802, by Alexander von Humboldt. This led to its mining and that of Chilean nitrate and to its application to soil in the United States and Europe after 1840.

The work of Liebig was a revolution for agriculture, and so other investigators started experimentation based on it. In England John Bennet Lawes and Joseph Henry Gilbert worked in the Rothamsted Experimental Station, founded by the former, and discovered that plants took nitrogen from the soil, and that salts needed to be in an available state to be absorbed by plants. Their investigations also produced the "superphosphate", consisting in the acid treatment of phosphate rock.This led to the invention and use of salts of potassium (K) and nitrogen (N) as fertilizers. Ammonia generated by the production of coke was recovered and used as fertiliser. Finally, the chemical basis of nutrients delivered to the soil in manure was understood and in the mid-19th century chemical fertilisers were applied. However, the dynamic interaction of soil and its life forms awaited discovery.

In 1856 J. T. Way discovered that ammonia contained in fertilisers was transformed into nitrates, and twenty years later R. W. Warington proved that this transformation was done by living organisms. In 1890 Sergei Winogradsky announced he had found the bacteria responsible for this transformation.

It was known that certain legumes could take up nitrogen from the air and fix it to the soil but it took the development of bacteriology towards the end of the 19th century to lead to an understanding of the role played in nitrogen fixation by bacteria. The symbiosis of bacteria and leguminous roots, and the fixation of nitrogen by the bacteria, were simultaneously discovered by German agronomist Hermann Hellriegel and Dutch microbiologist Martinus Beijerinck.

Crop rotation, mechanisation, chemical and natural fertilisers led to a doubling of wheat yields in Western Europe between 1800 and 1900.

Studies Concerning Soil Formation

The scientists who studied the soil in connection with agricultural practices had considered it mainly as a static substrate. However, soil is the result of evolution from more ancient geological materials. After studies of the improvement of the soil commenced, others began to study soil genesis and as a result also soil types and classifications.

In 1860, in Mississippi, Eugene W. Hilgard studied the relationship among rock material, climate, and vegetation, and the type of soils that were developed. He realised that the soils were dynamic, and considered soil types classification. Unfortunately his work was not continued. At the same time Vasily Dokuchaev (about 1870) was leading a team of soil scientists in Russia who conducted an extensive survey of soils, finding that similar basic rocks, climate and vegetation types lead to similar soil layering and types, and established the concepts for soil classifications. Due to the language barriers, the work of this team was not communicated to Western Europe until 1914 by a publication in German by K. D. Glinka, a member of the Russian team.

Curtis F. Marbut was influenced by the work of the Russian team, translated Glinka's publication into English, and as he was placed in charge of the U. S. National Cooperative Soil Survey, applied it to a national soil classification system.

Soil-Forming Processes

Soil formation, or pedogenesis, is the combined effect of physical, chemical, biological and anthropogenic processes working on soil parent material. Soil is said to be formed when organic matter has accumulated and colloids are washed downward, leaving deposits of clay, humus, iron oxide, carbonate, and gypsum, producing a distinct layer called the B horizon. This is a somewhat arbitrary definition as mixtures of sand, silt, clay and humus will support biological and agricultural activity before that time. These constituents are moved from one level to another by water and animal activity. As a result, layers (horizons) form in the soil profile. The alteration and movement of materials within a soil causes the formation of distinctive soil horizons.

How soil formation proceeds is influenced by at least five classic factors that are intertwined in the evolution of a soil. They are: parent material, climate, topography (relief), organisms, and time. When reordered to climate, relief, organisms, parent material, and time, they form the acronym CROPT.

An example of the development of a soil would begin with the weathering of lava flow bedrock, which would produce the purely mineral-based parent material from which the soil texture forms. Soil development would proceed most rapidly from bare rock of recent flows in a warm climate, under heavy and frequent rainfall. Under such conditions, plants become established very quickly on basaltic lava, even though there is very little organic material. The plants are supported by the porous rock as it is filled with nutrient-bearing water that carries dissolved minerals from the rocks and guano. Crevasses and pockets, local topography of the rocks, would hold fine materials and harbour plant roots. The developing plant roots are associated with mycorrhizal fungi that assist in breaking up the porous lava, and by these means organic matter and a finer mineral soil accumulate with time.

Parent Material

The mineral material from which a soil forms is called parent material. Rock, whether its origin is igneous, sedimentary, or metamorphic, is the source of all soil mineral materials and the origin of all plant nutrients with the exceptions of nitrogen, hydrogen and carbon. As the parent material is chemically and physically weathered, transported, deposited and precipitated, it is transformed into a soil.

Typical soil parent mineral materials are:

- Quartz: SiO_2
- Calcite: $CaCO_3$
- Feldspar: $KAlSi_3O_8$
- Mica (biotite): $K(Mg,Fe)_3AlSi_3O_{10}(OH)_2$

Classification of Parent Material

Loess field in Germany.

Parent materials are classified according to how they came to be deposited. Residual materials are mineral materials that have weathered in place from primary bedrock. Transported materials are those that have been deposited by water, wind, ice or gravity. Cumulose material is organic matter that has grown and accumulates in place.

Residual soils are soils that develop from their underlying parent rocks and have the same general chemistry as those rocks. The soils found on mesas, plateaux, and plains are residual soils. In the United States as little as three percent of the soils are residual.

Most soils derive from transported materials that have been moved many miles by wind, water, ice and gravity.

- Aeolian processes (movement by wind) are capable of moving silt and fine sand many hundreds of miles, forming loess soils (60–90 percent silt), common in the Midwest of North America and in Central Asia. Clay is seldom moved by wind as it forms stable aggregates.

- Water-transported materials are classed as either alluvial, lacustrine, or marine. Alluvial materials are those moved and deposited by flowing water. Sedimentary deposits settled in lakes are called lacustrine. Lake Bonneville and many soils around the Great Lakes of the United States are examples. Marine deposits, such as soils along the Atlantic and Gulf Coasts and in the Imperial Valley of California of the United States, are the beds of ancient seas that have been revealed as the land uplifted.
- Ice moves parent material and makes deposits in the form of terminal and lateral moraines in the case of stationary glaciers. Retreating glaciers leave smoother ground moraines and in all cases, outwash plains are left as alluvial deposits are moved downstream from the glacier.
- Parent material moved by gravity is obvious at the base of steep slopes as talus cones and is called colluvial material.

Cumulose parent material is not moved but originates from deposited organic material. This includes peat and muck soils and results from preservation of plant residues by the low oxygen content of a high water table. While peat may form sterile soils, muck soils may be very fertile.

Weathering of Parent Material

The weathering of parent material takes the form of physical weathering (disintegration), chemical weathering (decomposition) and chemical transformation. Generally, minerals that are formed under the high temperatures and pressures at great depths within the Earth's mantle are less resistant to weathering, while minerals formed at low temperature and pressure environment of the surface are more resistant to weathering. Weathering is usually confined to the top few meters of geologic material, because physical, chemical, and biological stresses generally decrease with depth. Physical disintegration begins as rocks that have solidified deep in the Earth are exposed to lower pressure near the surface and swell and become mechanically unstable. Chemical decomposition is a function of mineral solubility, the rate of which doubles with each 10 °C rise in temperature, but is strongly dependent on water to effect chemical changes. Rocks that will decompose in a few years in tropical climates will remain unaltered for millennia in deserts. Structural changes are the result of hydration, oxidation, and reduction.

- **Physical disintegration** is the first stage in the transformation of parent material into soil. Temperature fluctuations cause expansion and contraction of the rock, splitting it along lines of weakness. Water may then enter the cracks and freeze and cause the physical splitting of material along a path toward the center of the rock, while temperature gradients within the rock can cause exfoliation of "shells". Cycles of wetting and drying cause soil particles to be abraded to a finer size, as does the physical rubbing of material as it is moved by wind, water, and gravity. Water can deposit within rocks minerals that expand upon drying, thereby stressing the rock. Finally, organisms reduce parent material in size through the action of plant roots or digging on the part of animals.

- **Chemical decomposition** and **structural changes** result when minerals are made soluble by water or are changed in structure. The first three of the following list are solubility changes and the last three are structural changes.
- The **solution** of salts in water results from the action of bipolar water on ionic salt compounds producing a solution of ions and water, removing those minerals and reducing the rock's integrity.
- **Hydrolysis** is the transformation of minerals into polar molecules by the splitting of intervening water. This results in soluble acid-base pairs. For example, the hydrolysis of orthoclase-feldspar transforms it to acid silicate clay and basic potassium hydroxide, both of which are more soluble.
- In **carbonation**, the solution of carbon dioxide in water forms carbonic acid. Carbonic acid will transform calcite into more soluble calcium bicarbonate.
- **Hydration** is the inclusion of water in a mineral structure, causing it to swell and leaving it more stressed and easily decomposed.
- **Oxidation** of a mineral compound is the inclusion of oxygen in a mineral, causing it to increase its oxidation number and swell due to the relatively large size of oxygen, leaving it stressed and more easily attacked by water (hydrolysis) or carbonic acid (carbonation).
- **Reduction** the opposite of oxidation, means the removal of oxygen, hence the oxidation number of some part of the mineral is reduced, which occurs when oxygen is scarce. The reduction of minerals leaves them electrically unstable, more soluble and internally stressed and easily decomposed.

Of the above, hydrolysis and carbonation are the most effective.

Saprolite is a particular example of a residual soil formed from the transformation of granite, metamorphic and other types of bedrock into clay minerals. Often called "weathered granite", saprolite is the result of weathering processes that include: hydrolysis, chelation from organic compounds, hydration (the solution of minerals in water with resulting cation and anion pairs) and physical processes that include freezing and thawing. The mineralogical and chemical composition of the primary bedrock material, its physical features, including grain size and degree of consolidation, and the rate and type of weathering transforms the parent material into a different mineral. The texture, pH and mineral constituents of saprolite are inherited from its parent material.

Climate

The principal climatic variables influencing soil formation are effective precipitation (i.e., precipitation minus evapotranspiration) and temperature, both of which affect the rates of chemical, physical, and biological processes. The temperature and moisture both influence the organic matter content of soil through their effects on the balance between plant growth and microbial decomposition. Climate is the dominant factor in soil formation, and soils show the distinctive characteristics of the climate zones in which they form. For every 10 °C rise in temperature, the rates of biochemical reactions more than double. Mineral precipitation and temperature are the primary climatic influences on soil formation. If warm temperatures and abundant water are present in the profile at the same time, the processes of weathering, leaching, and plant growth will be maximized. Humid climates favor the growth of trees. In contrast, grasses are the dominant native vegetation in subhumid and semiarid regions, while shrubs and brush of various kinds dominate in arid areas.

Water is essential for all the major chemical weathering reactions. To be effective in soil formation, water must penetrate the regolith. The seasonal rainfall distribution, evaporative losses, site topography, and soil permeability interact to determine how effectively precipitation can influence soil formation. The greater the depth of water penetration, the greater the depth of weathering of the soil and its development. Surplus water percolating through the soil profile transports soluble and suspended materials from the upper to the lower layers. It may also carry away soluble materials in the surface drainage waters. Thus, percolating water stimulates weathering reactions and helps differentiate soil horizons. Likewise, a deficiency of water is a major factor in determining the characteristics of soils of dry regions. Soluble salts are not leached from these soils, and in some cases they build up to levels that curtail plant growth. Soil profiles in arid and semi-arid regions are also apt to accumulate carbonates and certain types of expansive clays.

The direct influences of climate include:

- A shallow accumulation of lime in low rainfall areas as caliche
- Formation of acid soils in humid areas
- Erosion of soils on steep hillsides
- Deposition of eroded materials downstream
- Very intense chemical weathering, leaching, and erosion in warm and humid regions where soil does not freeze

Climate directly affects the rate of weathering and leaching. Wind moves sand and smaller particles, especially in arid regions where there is little plant cover. The type and amount of precipitation influence soil formation by affecting the movement of ions and particles through the soil, and aid in the development of different soil profiles. Soil profiles are more distinct in wet and cool climates, where organic materials may accumulate, than in wet and warm climates, where organic materials are rapidly consumed. The effectiveness of water in weathering parent rock material depends on seasonal and daily temperature fluctuations. Cycles of freezing and thawing constitute an effective mechanism which breaks up rocks and other consolidated materials.

Climate also indirectly influences soil formation through the effects of vegetation cover and biological activity, which modify the rates of chemical reactions in the soil.

Topography

The topography, or relief, is characterized by the inclination (slope), elevation, and orientation of the terrain. Topography determines the rate of precipitation or runoff and rate of formation or erosion of the surface soil profile. The topographical setting may either hasten or retard the work of climatic forces.

Steep slopes encourage rapid soil loss by erosion and allow less rainfall to enter the soil before running off and hence, little mineral deposition in lower profiles. In semiarid regions, the lower effective rainfall on steeper slopes also results in less complete vegetative cover, so there is less plant contribution to soil formation. For all of these reasons, steep slopes prevent the formation of soil from getting very far ahead of soil destruction. Therefore, soils on steep terrain tend to have rather shallow, poorly developed profiles in comparison to soils on nearby, more level sites.

In swales and depressions where runoff water tends to concentrate, the regolith is usually more deeply weathered and soil profile development is more advanced. However, in the lowest landscape positions, water may saturate the regolith to such a degree that drainage and aeration are restricted. Here, the weathering of some minerals and the decomposition of organic matter are retarded, while the loss of iron and manganese is accelerated. In such low-lying topography, special profile features characteristic of wetland soils may develop. Depressions allow the accumulation of water, minerals and organic matter and in the extreme, the resulting soils will be saline marshes or peat bogs. Intermediate topography affords the best conditions for the formation of an agriculturally productive soil.

Organisms

Soil is the most abundant ecosystem on Earth, but the vast majority of organisms in soil are microbes, a great many of which have not been described. There may be a population limit of around one billion cells per gram of soil, but estimates of the number of species vary widely from 50,000 per gram to over a million per gram of soil. The total number of organisms and species can vary widely according to soil type, location, and depth.

Plants, animals, fungi, bacteria and humans affect soil formation (see soil biomantle and stonelayer). Animals, soil mesofauna and micro-organisms mix soils as they form burrows and pores, allowing moisture and gases to move about. In the same way, plant roots open channels in soils. Plants with deep taproots can penetrate many metres through the different soil layers to bring up nutrients from deeper in the profile. Plants with fibrous roots that spread out near the soil surface have roots that are easily decomposed, adding organic matter. Micro-organisms, including fungi and bacteria, effect chemical exchanges between roots and soil and act as a reserve of nutrients.

Humans impact soil formation by removing vegetation cover with erosion as the result. Their tillage also mixes the different soil layers, restarting the soil formation process as less weathered material is mixed with the more developed upper layers.

Earthworms, ants and termites mix the soil as they burrow, significantly affecting soil formation. Earthworms ingest soil particles and organic residues, enhancing the availability of plant nutrients in the material that passes through their bodies. They aerate and stir the soil and increase the stability of soil aggregates, thereby assuring ready infiltration of water. In addition, as ants and termites build mounds, they transport soil materials from one horizon to another.

In general, the mixing of the soil by the activities of animals, sometimes called pedoturbation, tends to undo or counteract the tendency of other soil-forming processes that create distinct horizons. Termites and ants may also retard soil profile development by denuding large areas of soil around their nests, leading to increased loss of soil by erosion. Large animals such as gophers, moles, and prairie dogs bore into the lower soil horizons, bringing materials to the surface. Their tunnels are often open to the surface, encouraging the movement of water and air into the subsurface layers. In localized areas, they enhance mixing of the lower and upper horizons by creating, and later refilling, underground tunnels. Old animal burrows in the lower horizons often become filled with soil material from the overlying A horizon, creating profile features known as crotovinas.

Vegetation impacts soils in numerous ways. It can prevent erosion caused by excessive rain that might result from surface runoff. Plants shade soils, keeping them cooler and slow evaporation of soil moisture, or conversely, by way of transpiration, plants can cause soils to lose moisture. Plants can form new chemicals that can break down minerals and improve the soil structure. The type and amount of vegetation depends on climate, topography, soil characteristics, and biological factors. Soil factors such as density, depth, chemistry, pH, temperature and moisture greatly affect the type of plants that can grow in a given location. Dead plants and fallen leaves and stems begin their decomposition on the surface. There, organisms feed on them and mix the organic material with the upper soil layers; these added organic compounds become part of the soil formation process.

Human activities widely influence soil formation. For example, it is believed that Native Americans regularly set fires to maintain several large areas of prairie grasslands in Indiana and Michigan. In more recent times, human destruction of natural vegetation and subsequent tillage of the soil for crop production has abruptly modified soil formation. Likewise, irrigating an arid region of soil drastically influences the soil-forming factors, as does adding fertilizer and lime to soils of low fertility.

Time

Time is a factor in the interactions of all the above. While a mixture of sand, silt and clay constitute the texture of a soil and the aggregation of those components produces peds, the development of a distinct B horizon marks the development of a soil. With time, soils will evolve features that depend on the interplay of the prior listed soil-forming factors. It takes decades to several thousand years for a soil to develop a profile. That time period depends strongly on climate, parent material, relief, and biotic activity. For example, recently deposited material from a flood exhibits no soil development as there has not been enough time for the material to form a structure that further defines soil. The original soil surface is buried, and the formation process must begin anew for this deposit. Over time the soil will develop a profile that depends on the intensities of biota and climate. While a soil can achieve relative stability of its properties for extended periods, the soil life cycle ultimately ends in soil conditions that leave it vulnerable to erosion. Despite the inevitability of soil retrogression and degradation, most soil cycles are long.

Soil-forming factors continue to affect soils during their existence, even on "stable" landscapes that are long-enduring, some for millions of years. Materials are deposited on top or are blown or washed from the surface. With additions, removals and alterations, soils are always subject to new conditions. Whether these are slow or rapid changes depends on climate, topography and biological activity.

Physical Properties of Soils

The physical properties of soils, in order of decreasing importance, are texture, structure, density, porosity, consistency, temperature, colour and resistivity. Soil texture is determined by the relative proportion of the three kinds of soil mineral particles, called soil separates: sand, silt, and clay. At the next larger scale, soil structures called peds are created from the soil separates when iron oxides, carbonates, clay, silica and humus, coat particles and cause them to adhere into larger, relatively stable secondary structures. Soil density, particularly bulk density, is a measure of soil compaction. Soil porosity consists of the void part of the soil volume and is occupied by gases or water. Soil consistency is the ability of soil to stick together. Soil temperature and colour are self-defining. Resistivity refers to the resistance to conduction of electric currents and affects the rate of corrosion of metal and concrete structures. These properties may vary through the depth of a soil profile. Most of these properties determine the aeration of the soil and the ability of water to infiltrate and to be held within the soil.

Influence of Soil Texture Separates on Some Properties of Soils

Property/behavior	Sand	Silt	Clay
Water-holding capacity	Low	Medium to high	High
Aeration	Good	Medium	Poor
Drainage rate	High	Slow to medium	Very slow
Soil organic matter level	Low	Medium to high	High to medium
Decomposition of organic matter	Rapid	Medium	Slow
Warm-up in spring	Rapid	Moderate	Slow
Compactability	Low	Medium	High
Susceptibility to wind erosion	Moderate (High if fine sand)	High	Low
Susceptibility to water erosion	Low (unless fine sand)	High	Low if aggregated, otherwise high
Shrink/Swell Potential	Very Low	Low	Moderate to very high
Sealing of ponds, dams, and landfills	Poor	Poor	Good
Suitability for tillage after rain	Good	Medium	Poor
Pollutant leaching potential	High	Medium	Low (unless cracked)
Ability to store plant nutrients	Poor	Medium to High	High
Resistance to pH change	Low	Medium	High

Texture

The mineral components of soil are sand, silt and clay, and their relative proportions determine a soil's texture. Properties that are influenced by soil texture, include porosity, permeability, infiltration, shrink-swell rate, water-holding capacity, and susceptibility to erosion. In the illustrated USDA textural classification triangle, the only soil in which neither sand, silt nor clay predominates is called "loam". While even pure sand, silt or clay may be considered a soil, from the perspective of food production a loam soil with a small amount of organic material is considered ideal. The mineral constituents of a loam soil might be 40% sand, 40%

silt and the balance 20% clay by weight. Soil texture affects soil behaviour, in particular its retention capacity for nutrients and water.

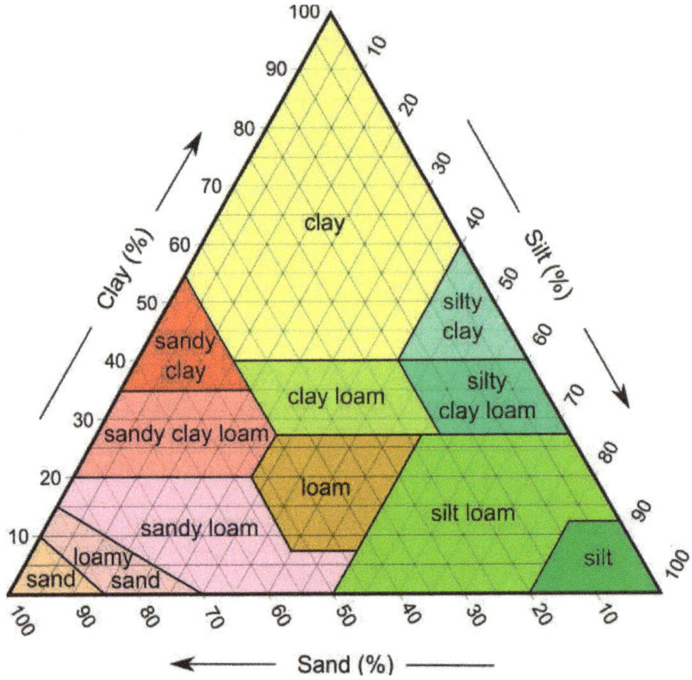

Soil types by clay, silt and sand composition as used by the USDA

Iron-rich soil near Paint Pots in Kootenay National Park, Canada

Sand and silt are the products of physical and chemical weathering of the parent rock; clay, on the other hand, is a most often the product of the precipitation of the dissolved parent rock as a secondary mineral. It is the surface area to volume ratio (specific surface area) of soil particles and the unbalanced ionic charges within those that determine their role in the fertility of soil, as measured by its cation exchange capacity. Sand is least active, followed by silt; clay is the most active. Sand's greatest benefit to soil is that it resists compaction and increases a soil's porosity. Silt is mineralogically like sand but with its higher specific surface area it is more chemically active than sand. But it is the clay content of soil, with its very high specific surface area and generally large number of negative charges, that gives a soil its high retention capacity for water and nutrients. Clay soils also resist wind and water erosion better than silty and sandy soils, as the particles bond tightly to each other.

Sand is the most stable of the mineral components of soil; it consists of rock fragments, primarily quartz particles, ranging in size from 2.0 to 0.05 mm (0.0787 to 0.0020 in) in diameter. Silt ranges in size from 0.05 to 0.002 mm (0.002 to 0.00008 in). Clay cannot be resolved by optical microscopes as its particles are 0.002 mm (7.9×10^{-5} in) or less in diameter and a thickness of only 10 angstroms (10^{-10} m). In medium-textured soils, clay is often washed downward through the soil profile and accumulates in the subsoil.

Soil components larger than 2.0 mm (0.079 in) are classed as rock and gravel and are removed before determining the percentages of the remaining components and the texture class of the soil, but are included in the name. For example, a sandy loam soil with 20% gravel would be called gravelly sandy loam.

When the organic component of a soil is substantial, the soil is called organic soil rather than mineral soil. A soil is called organic if:

- Mineral fraction is 0% clay and organic matter is 20% or more
- Mineral fraction is 0% to 50% clay and organic matter is between 20% and 30%
- Mineral fraction is 50% or more clay and organic matter 30% or more.

Structure

The clumping of the soil textural components of sand, silt and clay causes aggregates to form and the further association of those aggregates into larger units creates soil structures called peds (a contraction of the word pedolith). The adhesion of the soil textural components by organic substances, iron oxides, carbonates, clays, and silica, and the breakage of those aggregates from expansion-contraction, caused by freezing-thawing and wetting-drying cycles, shape soil into distinct geometric forms. The peds evolve into units which have various shapes, sizes and degrees of development. A soil clod, however, is not a ped but rather a mass of soil that results from mechanical disturbance of the soil. Soil structure affects aeration, water movement, conduction of heat, plant root growth and resistance to erosion. Water, in turn, has its strongest effect on soil structure due to its solution and precipitation of minerals and its effect on plant growth.

Soil structure often gives clues to its texture, organic matter content, biological activity, past soil evolution, human use, and the chemical and mineralogical conditions under which the soil formed. While texture is defined by the mineral component of a soil and is an innate property of the soil that does not change with agricultural activities, soil structure can be improved or destroyed by the choice and timing of farming practices.

Soil structural classes:

- Types: Shape and arrangement of peds
    - Platy: Peds are flattened one atop the other 1–10 mm thick. Found in the A-horizon of forest soils and lake sedimentation.
    - Prismatic and Columnar: Prismlike peds are long in the vertical dimension, 10–100 mm wide. Prismatic peds have flat tops, columnar peds have rounded tops. Tend to form in the B-horizon in high sodium soil where clay has accumulated.
    - Angular and subangular: Blocky peds are imperfect cubes, 5–50 mm, angular have sharp edges, subangular have rounded edges. Tend to form in the B-horizon where clay has accumulated and indicate poor water penetration.
    - Granular and Crumb: Spheroid peds of polyhedrons, 1–10 mm, often found in the A-horizon in the presence of organic material. Crumb peds are more porous and are considered ideal.
- Classes: Size of peds whose ranges depend upon the above type
    - Very fine or very thin: <1 mm platy and spherical; <5 mm blocky; <10 mm prismlike.
    - Fine or thin: 1–2 mm platy, and spherical; 5–10 mm blocky; 10–20 mm prismlike.
    - Medium: 2–5 mm platy, granular; 10–20 mm blocky; 20-50 prismlike.
    - Coarse or thick: 5–10 mm platy, granular; 20–50 mm blocky; 50–100 mm prismlike.
    - Very coarse or very thick: >10 mm platy, granular; >50 mm blocky; >100 mm prismlike.
- Grades: Is a measure of the degree of development or cementation within the peds that results in their strength and stability.
    - Weak: Weak cementation allows peds to fall apart into the three textural constituents, sand, silt and clay.
    - Moderate: Peds are not distinct in undisturbed soil but when removed they break into aggregates, some broken aggregates and little unaggregated material. This is considered ideal.
    - Strong:Peds are distinct before removed from the profile and do not break apart easily.
    - Structureless: Soil is entirely cemented together in one great mass such as slabs of clay or no cementation at all such as with sand.

At the largest scale, the forces that shape a soil's structure result from swelling and shrinkage that initially tend to act horizontally, causing vertically oriented prismatic peds.

Clayey soil, due to its differential drying rate with respect to the surface, will induce horizontal cracks, reducing columns to blocky peds. Roots, rodents, worms, and freezing-thawing cycles further break the peds into a spherical shape.

At a smaller scale, plant roots extend into voids and remove water causing the open spaces to increase, thereby decreasing aggregate size. At the same time, roots, fungal hyphae, and earthworms create microscopic tunnels that break up peds.

At an even smaller scale, soil aggregation continues as bacteria and fungi exude sticky polysaccharides which bind soil into smaller peds. The addition of the raw organic matter that bacteria and fungi feed upon encourages the formation of this desirable soil structure.

At the lowest scale, the soil chemistry affects the aggregation or dispersal of soil particles. The clay particles contain polyvalent cations which give the faces of clay layers localized negative charges. At the same time, the edges of the clay plates have a slight positive charge, thereby allowing the edges to adhere to the negative charges on the faces of other clay particles or to flocculate (form clumps). On the other hand, when monovalent ions, such as sodium, invade and displace the polyvalent cations, they weaken the positive charges on the edges, while the negative surface charges are relatively strengthened. This leaves negative charge on the clay faces that repel other clay, causing the particles to push apart, and by doing,the flocculation of clay particles into larger, open assemblages. As a result, the clay disperses and settles into voids between peds, causing those to close. In this way the open structure of the soil is destroyed and the soil is made impenetrable to air and water. Such sodic soil tends to form columnar structures near the surface.

Density

Soil particle density is typically 2.60 to 2.75 grams per cm^3 and is usually unchanging for a given soil. Soil particle density is lower for soils with high organic matter content, and is higher for soils with high iron-oxides content. Soil bulk density is equal to the dry mass of the soil divided by the volume of the soil; i.e., it includes air space and organic materials of the soil volume. The soil bulk density of cultivated loam is about 1.1 to 1.4 g/cm^3 (for comparison water is 1.0 g/cm^3).Soil bulk density is highly variable for a given soil. A lower bulk density by itself does not indicate suitability for plant growth due to the influence of soil texture and structure. A high bulk density is indicative of either soil compaction or high sand content. Soil bulk density is inherently always less than the soil particle density.

Representative bulk densities of soils. The percentage pore space was calculated using 2.7 g/cm3 for particle density except for the peat soil, which is estimated.

Soil treatment and identification	**Bulk density g/cm^3**	**Pore space %**
Tilled surface soil of a cotton field	1.3	51
Trafficked inter-rows where wheels passed surface	1.67	37
Traffic pan at 25 cm deep	1.7	36
Undisturbed soil below traffic pan, clay loam	1.5	43

Rocky silt loam soil under aspen forest	1.62	40
Loamy sand surface soil	1.5	43
Decomposed peat	0.55	65

Porosity

Pore space is that part of the bulk volume of soil that is not occupied by either mineral or organic matter but is open space occupied by either gases or water. In a productive, medium-textured soil the total pore space is typically about 50% of the soil volume. Pore size varies considerably; the smallest pores (cryptopores; <0.1 μm) hold water too tightly for use by plant roots; plant-available water is held in ultramicropores, micropores and mesopores (0.1-75 μm); and macropores (>75 μm) are generally air-filled when the soil is at field capacity.

Soil texture determines total volume of the smallest pores; clay soils have smaller pores, but more total pore space than sands. Soil structure has a strong influence on the larger pores that affect soil aeration, water infiltration and drainage. Tillage has the short-term benefit of temporarily increasing the number of pores of largest size, but these can be rapidly degraded by the destruction of soil aggregation.

The pore size distribution affects the ability of plants and other organisms to access water and oxygen; large, continuous pores allow rapid transmission of air, water and dissolved nutrients through soil, and small pores store water between rainfall or irrigation events. Pore size variation also compartmentalizes the soil pore space such that many microorganisms are not in direct competition with one another, which may explain not only the large number of species present, but the fact that functionally redundant microorganisms (organisms with the same ecological niche) can co-exist within the same soil.

Consistency

Consistency is the ability of soil to stick to itself or to other objects (cohesion and adhesion respectively) and its ability to resist deformation and rupture. It is of approximate use in predicting cultivation problems and the engineering of foundations. Consistency is measured at three moisture conditions: air-dry, moist, and wet. In those conditions the consistency quality depends upon the clay content. In the wet state, the two qualities of stickiness and plasticity are assessed. A soil's resistance to fragmentation and crumbling is assessed in the dry state by rubbing the sample. Its resistance to shearing forces is assessed in the moist state by thumb and finger pressure. Additionally, the cemented consistency depends on cementation by substances other than clay, such as calcium carbonate, silica, oxides and salts; moisture content has little effect on its assessment. The measures of consistency border on subjective compared to other measures such as pH, since they employ the apparent feel of the soil in those states.

The terms used to describe the soil consistency in three moisture states and a last not affected by the amount of moisture are as follows:

- Consistency of Dry Soil: loose, soft, slightly hard, hard, very hard, extremely hard

- Consistency of Moist Soil: loose, very friable, friable, firm, very firm, extremely firm
- Consistency of Wet Soil: nonsticky, slightly sticky, sticky, very sticky; nonplastic, slightly plastic, plastic, very plastic
- Consistency of Cemented Soil: weakly cemented, strongly cemented, indurated (requires hammer blows to break up)

Soil consistency is useful in estimating the ability of soil to support buildings and roads. More precise measures of soil strength are often made prior to construction.

Temperature

Soil temperature depends on the ratio of the energy absorbed to that lost. Soil has a temperature range between -20 to 60 °C. Soil temperature regulates seed germination, plant and root growth and the availability of nutrients. Below 50 cm (20 in), soil temperature seldom changes and can be approximated by adding 1.8 °C (2 °F) to the mean annual air temperature. Soil temperature has important seasonal, monthly and daily variations. Fluctuations in soil temperature are much lower with increasing soil depth. Heavy mulching (a type of soil cover) can slow the warming of soil, and, at the same time, reduce fluctuations in surface temperature.

Most often, agricultural activities must adapt to soil temperatures by:

- maximizing germination and growth by timing of planting
- optimizing use of anhydrous ammonia by applying to soil below 10 °C (50 °F)
- preventing heaving and thawing due to frosts from damaging shallow-rooted crops
- preventing damage to desirable soil structure by freezing of saturated soils
- improving uptake of phosphorus by plants

Soil temperatures can be raised by drying soils or the use of clear plastic mulches. Organic mulches slow the warming of the soil.

There are various factors that affect soil temperature, such as water content, soil color, and relief (slope, orientation, and elevation), and soil cover (shading and insulation). The color of the ground cover and its insulating properties have a strong influence on soil temperature. Whiter soil tends to have a higher albedo than blacker soil cover, which encourages whiter soils to have lower soil temperatures. The specific heat of soil is the energy required to raise the temperature of soil by 1 °C. The specific heat of soil increases as water content increases, since the heat capacity of water is greater than that of dry soil. The specific heat of pure water is ~ 1 calorie per gram, the specific heat of dry soil is ~ 0.2 calories per gram, hence, the specific heat of wet soil is ~ 0.2 to 1 calories per gram. Also, a tremendous energy (~540 cal/g) is required to evaporate water (known as the heat of vaporization). As such, wet soil usually warms more slowly than dry soil - wet surface soil is typically 3 to 6 °C colder than dry surface soil.

Soil heat flux refers to the rate at which heat energy moves through the soil in response to a temperature difference between two points in the soil. The heat flux density is the amount of energy that flows through soil per unit area per unit time and has both magnitude and direction. For the simple case of conduction into or out of the soil in the vertical direction, which is most often applicable the heat flux density is:

$$q_x = -k\frac{\delta T}{\delta x}$$

In SI units

q is the heat flux density, in SI the units are W·m^{-2}

k is the soils' conductivity, W·m^{-1}·K^{-1}. The thermal conductivity is sometimes a constant, otherwise an average value of conductivity for the soil condition between the surface and the point at depth is used.

δT is the temperature difference (temperature gradient) between the two points in the soil between which the heat flux density is to be calculated. In SI the units are kelvin, K.

δx is the distance between the two points within the soil, at which the temperatures are measured and between which the heat flux density is being calculated. In SI the units are meters m, and where x is measured positive downward.

Heat flux is in the direction opposite the temperature gradient, hence the minus sign. That is to say, if the temperature of the surface is higher than at depth x the negative sign will result in a positive value for the heat flux q, and which is interpreted as the heat being conducted into the soil.

Component	Thermal Conductivity (W·m-1·K-1)
Quartz	8.8
Clay	2.9
Organic matter	0.25
Water	0.57
Ice	2.4
Air	0.025
Dry soil	0.2-0.4
Wet soil	1-3

Soil temperature is important for the survival and early growth of seedlings. Soil temperatures affect the anatomical and morphological character of root systems.All physical, chemical, and biological processes in soil and roots are affected in particular because of the increased viscosities of water and protoplasm at low temperatures. In general, climates that do not preclude survival and growth of white spruce above ground are sufficiently benign to provide soil temperatures able to maintain white spruce root systems. In some northwestern parts of the range, white spruce occurs on permafrost sites and although young unlignified roots of conifers may have little resistance to freezing, less than half of the "secondary mature" root system of white spruce was killed by exposure to a temperature of 23.3 °C in multiple year experiment with containerized trees from local nurseries in Massachusetts.

Optimum temperatures for tree root growth range between 10 °C and 25 °C in general and for spruce in particular. In 2-week-old white spruce seedlings that were then grown for 6 weeks in soil at temperatures of 15 °C, 19 °C, 23 °C, 27 °C, and 31 °C; shoot height, shoot dry weight, stem diameter, root penetration, root volume, and root dry weight all reached maxima at 19 °C.

However, whereas strong positive relationships between soil temperature (5 °C to 25 °C) and growth have been found in trembling aspen and balsam poplar, white and other spruce species have shown little or no changes in growth with increasing soil temperature.

Such insensitivity to soil low temperature may be common among a number of western and boreal conifers.

Color

Soil colour is often the first impression one has when viewing soil. Striking colours and contrasting patterns are especially noticeable. The Red River of the South carries sediment eroded from extensive reddish soils like Port Silt Loam in Oklahoma. The Yellow River in China carries yellow sediment from eroding loess soils. Mollisols in the Great Plains of North America are darkened and enriched by organic matter. Podsols in boreal forests have highly contrasting layers due to acidity and leaching.

In general, color is determined by the organic matter content, drainage conditions, and degree of oxidation. Soil color, while easily discerned, has little use in predicting soil characteristics. It is of use in distinguishing boundaries within a soil profile, determining the origin of a soil's parent material, as an indication of wetness and waterlogged conditions, and as a qualitative means of measuring organic, salt and carbonate contents of soils. Color is recorded in the Munsell color system as for instance 10YR3/4 *Dusky Red.*

Soil color is primarily influenced by soil mineralogy. Many soil colours are due to various iron minerals. The development and distribution of colour in a soil profile result from chemical and biological weathering, especially redox reactions. As the primary minerals in soil parent material weather, the elements combine into new and colourful compounds. Iron forms secondary minerals of a yellow or red colour, organic matter decomposes into black and brown compounds, and manganese, sulfur and nitrogen can form black mineral deposits. These pigments can produce various colour patterns within a soil. Aerobic conditions produce uniform or gradual colour changes, while reducing environments (anaerobic) result in rapid colour flow with complex, mottled patterns and points of colour concentration.

Resistivity

Soil resistivity is a measure of a soil's ability to retard the conduction of an electric current. The electrical resistivity of soil can affect the rate of galvanic corrosion of metallic structures in contact with the soil. Higher moisture content or increased electrolyte concentration can lower resistivity and increase conductivity, thereby increasing the rate of corrosion. Soil resistivity values typically range from about 2 to 1000 $\Omega \cdot m$, but more extreme values are not unusual.

Soil Water

Water that enters a field is removed from a field by runoff, drainage, evaporation or transpiration. Runoff is the water that flows on the surface to the edge of the field; drainage is the water that flows through the soil downward or toward the edge of the field underground; evaporative water loss from a field is that part of the water that evaporates into the atmosphere directly from the field's surface; transpiration is the loss of water from the field by its evaporation from the plant itself.

Water affects soil formation, structure, stability and erosion but is of primary concern with respect to plant growth. Water is essential to plants for four reasons:

- It constitutes 80%-95% of the plant's protoplasm.
- It is essential for photosynthesis.
- It is the solvent in which nutrients are carried to, into and throughout the plant.
- It provides the turgidity by which the plant keeps itself in proper position.

In addition, water alters the soil profile by dissolving and re-depositing minerals, often at lower levels, and possibly leaving the soil sterile in the case of extreme rainfall and drainage. In a loam soil, solids constitute half the volume, gas one-quarter of the volume, and water one-quarter of the volume of which only half of which will be available to most plants.

A flooded field will drain the gravitational water under the influence of gravity until water's adhesive and cohesive forces resist further drainage at which point it is said to have reached field capacity. At that point, plants must apply suction to draw water from a soil. The water that plants may draw from the soil is called the available water. Once the available water is used up the remaining moisture is called unavailable water as the plant cannot produce sufficient suction to draw that water in. A plant must produce suction that increases from zero for a flooded field to 1/3 bar at field dry condition (one bar is a little less than one atmosphere pressure). At 15 bar suction, wilting percent, seeds will not germinate, plants begin to wilt and then die. Water moves in soil under the influence of gravity, osmosis and capillarity. When water enters the soil, it displaces air from some of the pores, since air content of a soil is inversely related to its water content.

The rate at which a soil can absorb water depends on the soil and its other conditions. As a plant grows, its roots remove water from the largest pores first. Soon the larger pores hold only air, and the remaining water is found only in the intermediate- and smallest-sized pores. The water in the smallest pores is so strongly held to particle surfaces that plant roots cannot pull it away. Consequently, not all soil water is available to plants. When saturated, the soil may lose nutrients as the water drains. Water moves in a draining field under the influence of pressure where the soil is locally saturated and by capillarity pull to dryer parts of the soil. Most plant water needs are supplied from the suction caused by evaporation from plant leaves and 10% is supplied by "suction" created by osmotic pressure differences between the plant interior and the soil water. Plant roots must seek out water. Insufficient water will damage the yield of a crop. Most of the available water is used in transpiration to pull nutrients into the plant.

Water Retention Forces

Water is retained in a soil when the adhesive force of attraction that water's hydrogen atoms have for the oxygen of soil particles is stronger than the cohesive forces that water's hydrogen feels for other water oxygen atoms. When a field is flooded, the soil pore space is completely filled by water. The field will drain under the force of gravity until it reaches what is called field capacity, at which point the smallest pores are filled with water and the largest with water and gases.The total amount of water held when field capacity is reached is a function of the specific surface area of the soil particles. As a result, high clay and high organic soils have higher field capacities. The total force required to pull or push water out of soil is termed suction and usually expressed in units of bars (10^5 pascal) which is just a little less than one-atmosphere pressure. Alternatively, the terms "tension" or "moisture potential" may be used.

Moisture Classification

The forces with which water is held in soils determine its availability to plants. Forces of adhesion hold water strongly to mineral and humus surfaces and less strongly to itself by cohesive forces. A plant's root may penetrate a very small volume of water that is adhering to soil and be initially able to draw in water that is only lightly held by the cohesive forces. But as the droplet is drawn down, the forces of adhesion of the water for the soil particles make reducing the volume of water increasingly difficult until the plant cannot produce sufficient suction to use the remaining water. The remaining water is considered unavailable. The amount of available water depends upon the soil texture and humus amounts and the type of plant attempting to use the water. Cacti, for example, can produce greater suction than can agricultural crop plants.

The following description applies to a loam soil and agricultural crops. When a field is flooded, it is said to be saturated and all available air space is occupied by water. The suction required to draw water into a plant root is zero. As the field drains under the influence of gravity (drained water is called gravitational water or drain-able water), the suction a plant must produce to use such water increases to 1/3 bar. At that point, the soil is said to have reached field capacity, and plants that use the water must produce increasingly higher suction, finally up to 15 bar. At 15 bar suction, the soil water amount is called wilting percent. At that suction the plant cannot sustain its water needs as water is still being lost from the plant by transpiration; the plant's turgidity is lost, and it wilts. The next level, called air-dry, occurs at 1000 bar suction. Finally the oven dry condition is reached at 10,000 bar suction. All water below wilting percentage is called unavailable water.

Soil Moisture Content

When the soil moisture content is optimal for plant growth, the water in the large and intermediate size pores can move about in the soil and be easily used by plants. The amount of water remaining in a soil drained to field capacity and the amount that is available are functions of the soil type. Sandy soil will retain very little water, while clay will hold the maximum amount. The time required to drain a field from flooded condition for a clay loam that begins at 43% water by weight to a field capacity of 21.5% is six days, whereas a sand loam that is flooded to its maximum of 22% water will take two days to reach field capacity of 11.3% water. The available water for the clay loam might be 11.3% whereas for the sand loam it might be only 7.9% by weight.

Wilting point, field capacity, and available water capacity of various soil textures

Soil Texture	Wilting Point Water per foot of soil depth		Field Capacity Water per foot of soil depth		Available water capacity Water per foot of soil depth	
	%	in.	%	in.	%	in.
Medium sand	1.7	0.3	6.8	1.2	5.1	0.9
Fine sand	2.3	0.4	8.5	1.5	6.2	1.1
Sandy loam	3.4	0.6	11.3	2.0	7.9	1.4
Fine sandy loam	4.5	0.8	14.7	2.6	10.2	1.8
Loam	6.8	1.2	18.1	3.2	11.3	2.0
Silt loam	7.9	1.4	19.8	3.5	11.9	2.1
Clay loam	10.2	1.8	21.5	3.8	11.3	2.0
Clay	14.7	2.6	22.6	4.0	7.9	1.4

The above are average values for the soil textures as the percentages of sand, silt and clay vary.

Water Flow in Soils

Water moves through soil due to the force of gravity, osmosis and capillarity. At zero to one-third bar suction, water is pushed through soil from the point of its application under the force of gravity and the pressure gradient created by the pressure of the water; this is called saturated flow. At higher suction, water movement is pulled by capillarity from wetter toward dryer soil. This is caused by water's adhesion to soil solids, and is called unsaturated flow.

Water infiltration and movement in soil is controlled by six factors:

- Soil texture
- Soil structure. Fine-textured soils with granular structure are most favourable to infiltration of water.
- The amount of organic matter. Coarse matter is best and if on the surface helps prevent the destruction of soil structure and the creation of crusts.
- Depth of soil to impervious layers such as hardpans or bedrock
- The amount of water already in the soil
- Soil temperature. Warm soils take in water faster while frozen soils may not be able to absorb depending on the type of freezing.

Water infiltration rates range from 0.25 cm (0.098 in) per hour for high clay soils to 2.5 cm (0.98 in) per hour for sand and well stabilised and aggregated soil structures. Water flows through the ground unevenly, called "gravity fingers", because of the surface tension between water particles.Tree roots create paths for rainwater flow through soil by breaking though soil including clay layers: one study showed roots increasing infiltration of water by 153% and another study showed an increase by 27 times.Flooding temporarily increases soil permeability in river beds, helping to recharge aquifers.

Saturated Flow

Water applied to a soil is pushed by pressure gradients from the point of its application where it is saturated locally, to less saturated areas. Once soil is completely wetted, any more water will move downward, or percolate, carrying with it clay, humus and nutrients, primarily cations, out of the range of plant roots. In order of decreasing solubility, the leached nutrients are:

- Calcium
- Magnesium, Sulfur, Potassium; depending upon soil composition
- Nitrogen; usually little, unless nitrate fertiliser was applied recently
- Phosphorus; very little as its forms in soil are of low solubility.

In the United States percolation water due to rainfall ranges from zero inches just east of the Rocky Mountains to twenty or more inches in the Appalachian Mountains and the north coast of the Gulf of Mexico.

Unsaturated Flow

At suctions less than one-third bar, water moves in all directions via unsaturated flow at a rate that is dependent on the square of the diameter of the water-filled pores. Water is pulled by capillary action due to the adhesion force of water to the soil solids, producing a suction gradient from wet towards drier soil. Doubling the diameter of the pores increases the flow rate by a factor of four. Large pores drained by gravity and not filled with water do not greatly increase the flow rate for unsaturated flow. Water flow is primarily from coarse-textured soil into fine-textured soil and is slowest in fine-textured soils such as clay.

Water Uptake by Plants

Of equal importance to the storage and movement of water in soil is the means by which plants acquire it and their nutrients. Ninety percent of water is taken up by plants as passive absorption caused by the pulling force of water evaporating (transpiring) from the long column of water that leads from the plant's roots to its leaves. In addition, the high concentration of salts within plant roots creates an osmotic pressure gradient that pushes soil water into the roots. Osmotic absorption becomes more important during times of low water transpiration caused by lower temperatures (for example at night) or high humidity. It is the process that causes guttation.

Root extension is vital for plant survival. A study of a single winter rye plant grown for four months in one cubic foot of loam soil showed that the plant developed 13,800,000 roots, a total of 385 miles in length with 2,550 square feet in surface area; and 14 billion hair roots of 6,600 miles total length and 4,320 square feet total area; for a total surface area of 6,870 square feet (83 ft squared). The total surface area of the loam soil was estimated to be 560,000 square feet. In other words, the roots were in contact with only 1.2% of the soil.

Roots must seek out water as the unsaturated flow of water in soil can move only at a rate of up to 2.5 cm (one inch) per day; as a result they are constantly dying and growing as they seek out high concentrations of soil moisture. Insufficient soil moisture, to the point of causing wilting, will cause permanent damage and crop yields will suffer. When grain sorghum was exposed to soil suction as low as 13.0 bar during the seed head emergence through bloom and seed set stages of growth, its production was reduced by 34%.

Consumptive Use and Water Efficiency

Only a small fraction (0.1% to 1%) of the water used by a plant is held within the plant. The majority is ultimately lost via transpiration, while evaporation from the soil surface is also substantial. Transpiration plus evaporative soil moisture loss is called evapotranspiration. Evapotranspiration plus water held in the plant totals to consumptive use, which is nearly identical to evapotranspiration.

The total water used in an agricultural field includes runoff, drainage and consumptive use. The use of loose mulches will reduce evaporative losses for a period after a field is irrigated, but in the end the total evaporative loss will approach that of an uncovered soil. The benefit from mulch is to keep the moisture available during the seedling stage. Water use efficiency is measured by transpiration ratio, which is the ratio of the total water transpired by a plant to the dry weight of the harvested plant. Transpiration ratios for crops range from 300 to 700. For example, alfalfa may have a transpiration ratio of 500 and as a result 500 kilograms of water will produce one kilogram of dry alfalfa.

Soil Atmosphere

The atmosphere of soil is radically different from the atmosphere above. The consumption of oxygen, by microbes and plant roots and their release of carbon dioxide, decrease oxygen and increase carbon dioxide concentration. Atmospheric CO_2 concentration is 0.04%, but in the soil pore space it may range from 10 to 100 times that level. At extreme levels CO_2 is toxic. In addition, the soil voids are saturated with water vapour. Adequate porosity is necessary, not just to allow the penetration of water, but also to allow gases to diffuse in and out. Movement of gases is by diffusion from high concentrations to lower. Oxygen diffuses in and is consumed and excess levels of carbon dioxide, diffuse out with other gases as well as water. Soil texture and structure strongly affect soil porosity and gas diffusion. It is the total pore space (porosity) of soil not the pore size that determines the rate of diffusion of gases into and out of soil. A Platy soil structure and compacted soils (low porosity) impede gas flow, and a deficiency of oxygen may encourage anaerobic bacteria to reduce (strip oxygen) from nitrate NO_3 to the gases N_2, N_2O, and NO, which are then lost to the atmosphere, thereby depleting the soil of nitrogen. Aerated soil is also a net sink of methane CH_4 but a net producer of greenhouse gases when soils are depleted of oxygen and subject to elevated temperatures.

Composition of Soil Particles

Soil particles can be classified by their chemical composition (mineralogy) as well as their size. The particle size distribution of a soil, its texture, determines many of the properties of that soil, but the mineralogy of those particles can strongly modify those properties. The mineralogy of the finest soil particles, clay, is especially important.

Gravel, Sand and Silt

Gravel, sand and silt are the larger soil particles, and their mineralogy is often inherited from the parent material of the soil, but may include products of weathering (such as concretions of calcium carbonate or iron oxide), or residues of plant and animal life (such as silica phytoliths). Quartz is the most common mineral in the sand or silt fraction as it is resistant to chemical weathering; other common minerals are feldspars, micas and ferromagnesian minerals such as pyroxenes, amphiboles and olivines.

Mineral Colloids; Soil Clays

Due to its high specific surface area and its unbalanced negative charges, clay is the most active mineral component of soil. It is a colloidal and most often a crystalline material. In soils, clay is a soil textural class and is defined in a physical sense as any mineral particle less than 2 μm (8×10^{-5} in) in effective diameter. Many soil minerals, such as gypsum, carbonates, or quartz, are small enough to be classified as clay based on their physical size, but chemically they do not afford the same utility as do clay minerals. Chemically, clay is a range of minerals with certain reactive properties.

Clay was once thought to be very small particles of quartz, feldspar, mica, hornblende or augite, but it is now known to be (with the exception of mica-based clays) a precipitate with a mineralogical composition that is dependent on but different from its parent materials and is classed as a secondary mineral. The type of clay that is formed is a function of the parent material and the composition of the minerals in solution. Clay minerals continue to be formed as long as the soil exists.Mica-based clays result from a modification of the primary mica mineral in such a way that it behaves and is classed as a clay. Most clays are crystalline, but some are amorphous. The clays of a soil are a mixture of the various types of clay, but one type predominates.

There are four groups of clay: layer silicates; crystalline chain silicates; metal oxides and hydroxides and oxy-oxides; and amorphous; and allophanes. Most clays are crystalline and most are made up of three or four planes of oxygen held together by planes of aluminium and silicon by way of ionic bonds that together form a single layer of clay. The spatial arrangement of the oxygen atoms determines clay's structure. Half of the weight of clay is oxygen, but on a volume basis oxygen is ninety percent. The layers of clay are sometimes held together through hydrogen bonds or potassium bridges and as a result will swell less in the presence of water. Other clays, such as montmorillonite, have layers that are loosely attached and will swell greatly when water intervenes between the layers.

There are four groups of clays:

- Layer Crystalline alumino-silica clays: montmorillonite, illite, vermiculite, chlorite, kaolinite.
- Crystalline Chain carbonate and sulfate minerals: calcite ($CaCO_3$), dolomite ($CaMg(CO_3)_2$) and gypsum ($CaSO_4 \cdot 2H2O$).
- Amorphous clays: young mixtures of silica (SiO_2-OH) and alumina ($Al(OH)_3$) which have not had time to form regular crystals.

- Sesquioxide clays: old, highly leached clays which result in oxides of iron, aluminium and titanium.

Alumino-Silica Clays

Alumino-silica clays are characterised by their regular crystalline structure. Oxygen in ionic bonds with silicon forms a tetrahedral coordination (silicon at the center) which in turn forms sheets of silica. Two sheets of silica are bonded together by a plane of aluminium which forms an octahedral coordination, called alumina, with the oxygens of the silica sheet above and that below it. Hydroxyl ions (OH^-) sometimes substitute for oxygen. During the clay formation process, Al^{3+}may substitute for Si^{4+} in the silica layer, and as much as one fourth of the aluminium Al^{3+} may be substituted by Zn^{2+}, Mg^{2+} or Fe^{2+} in the alumina layer. The substitution of lower-valence cations for higher-valence cations (isomorphous substitution) gives clay a local negative charge on an oxygen atom that attracts and holds water and positively charged soil cations, some of which are of value for plant growth. Isomorphous substitution occurs during the clay's formation and does not change with time.

- **Montmorillonite** clay is made of four planes of oxygen with two silicon and one central aluminium plane intervening. The alumino-silicate montmorillonite clay is said to have a 2:1 ratio of silicon to aluminium. The seven planes together form a single crystal of montmorillonite. The crystals are weakly held together and water may intervene, causing the clay to swell up to ten times its dry volume. It occurs in soils which have had little leaching, hence it is found in arid regions. As the crystals are not bonded face to face, the entire surface is exposed and available for surface reactions, hence it has a high cation exchange capacity (CEC).
- **Illite** is a 2:1 clay similar in structure to montmorillonite but has potassium bridges between the faces of the clay crystals and the degree of swelling depends on the degree of weathering of the potassium. The active surface area is reduced due to the potassium bonds. Illite originates from the modification of mica, a primary mineral. It is often found together with montmorillonite and its primary minerals. It has moderate CEC.
- **Vermiculite** is a mica-based clay similar to illite, but the crystals of clay are held together more loosely by hydrated magnesium and it will swell, but not as much as does montmorillonite. It has very high CEC.
- **Chlorite** is similar to vermiculite, but the loose bonding by occasional hydrated magnesium, as in vermiculite, is replaced by a hydrated magnesium sheet, that firmly bonds the planes above and below it. It has two planes of silicon, one of aluminium and one of magnesium; hence it is a 2:2 clay. Chlorite does not swell and it has low CEC.
- **Kaolinite** is very common, highly weathered clay, and more common than montmorillonite in acid soils. It has one silica and one alumina plane per crystal; hence it is a 1:1 type clay. One plane of silica of montmorillonite is dissolved and is replaced with hydroxyls, which produces strong hydrogen bonds to the oxygen in the next crystal of clay. As a result, kaolinite does not swell in water

and has a low specific surface area, and as almost no isomorphous substitution has occurred it has a low CEC. Where rainfall is high, acid soils selectively leach more silica than alumina from the original clays, leaving kaolinite. Even heavier weathering results in sesquioxide clays.

silica-sesquioxide

Crystalline Chain Clays

The carbonate and sulfate minerals are much more soluble and hence are found primarily in desert soils where leaching is less active.

Amorphous Clays

Amorphous clays are young, and commonly found in volcanic ash. They are mixtures of alumina and silica which have not formed the ordered crystal shape of alumino-silica clays which time would provide. The majority of their negative charges originates from hydroxyl ions, which can gain or lose a hydrogen ion (H^+) in response to soil pH, in such way was as to buffer the soil pH. They may have either a negative charge provided by the attached hydroxyl ion (OH^-), which can attract a cation, or lose the hydrogen of the hydroxyl to solution and display a positive charge which can attract anions. As a result, they may display either high CEC in an acid soil solution, or high anion exchange capacity in a basic soil solution.

Sesquioxide Clays

Sesquioxide **clays** are a product of heavy rainfall that has leached most of the silica from alumino-silica clay, leaving the less soluble oxides iron hematite (Fe_2O_3), iron hydroxide ($Fe(OH)_3$), aluminium hydroxide gibbsite ($Al(OH)_3$), hydrated manganese birnessite (MnO_2). It takes hundreds of thousands of years of leaching to create sesquioxide clays. *Sesqui* is Latin for "one and one-half": there are three parts oxygen to two parts iron or aluminium; hence the ratio is one and one-half (not true for all). They are hydrated and act as either amorphous or crystalline. They are not sticky and do not swell, and soils high in them behave much like sand and can rapidly pass water. They are able to hold large quantities of phosphates. Sesquioxides have low CEC but are able to hold anions as well as cations. Such soils range from yellow to red in colour. Such clays tend to hold phosphorus so tightly that it is unavailable for absorption by plants.

Organic Colloids

Humus is the final state of decomposition of organic matter. While it may linger for a thousand years, on the larger scale of the age of the mineral soil components, it is temporary. It is composed of the very stable lignins (30%) and complex sugars (polyuronides, 30%), proteins (30%), waxes, and fats that are resistant to breakdown by microbes. Its chemical assay is 60% carbon, 5% nitrogen, some oxygen and the remainder hydrogen, sulfur, and phosphorus. On a dry weight basis, the CEC of humus is many times greater than that of clay.

Carbon and Terra Preta

In the extreme environment of high temperatures and the leaching caused by the heavy rain of tropical rain forests, the clay and organic colloids are largely destroyed. The heavy rains wash the alumino-silicate clays from the soil leaving only sesquioxide clays of low CEC. The high temperatures and humidity allow bacteria and fungi to virtually dissolve any organic matter on the rain-forest floor overnight and much of the nutrients are volatilized or leached from the soil and lost. However, carbon in the form of charcoal is far more stable than soil colloids and is capable of performing many of the functions of the soil colloids of sub-tropical soils. Soil containing substantial quantities of charcoal, of an anthropogenic origin, is called terra preta. Research into terra preta is still young but is promising. Fallow periods "on the Amazonian Dark Earths can be as short as 6 months, whereas fallow periods on oxisols are usually 8 to 10 years long"

Soil Chemistry

The chemistry of a soil determines its ability to supply available plant nutrients and affects its physical properties and the health of its microbial population. In addition, a soil's chemistry also determines its corrosivity, stability, and ability to absorb pollutants and to filter water. It is the surface chemistry of mineral and organic colloids that determines soil's chemical properties. "A colloid is a small, insoluble, nondiffusible particle larger than a molecule but small enough to remain suspended in a fluid medium without settling. Most soils contain organic colloidal particles called humus as well as the inorganic colloidal particles of clays." The very high specific surface area of colloids and their net charges, gives soil its ability to hold and release ions. Negatively charged sites on colloids attract and release cations in what is referred to as cation exchange. Cation-exchange capacity (CEC) is the amount of exchangeable cations per unit weight of dry soil and is expressed in terms of milliequivalents of positively charged ions per 100 grams of soil (or centimoles of positive charge per kilogram of soil; $cmol_c/kg$). Similarly, positively charged sites on colloids can attract and release anions in the soil giving the soil anion exchange capacity (AEC).

Cation and Anion Exchange

The cation exchange, that takes place between colloids and soil water, buffers (moderates) soil pH, alters soil structure, and purifies percolating water by adsorbing cations of all types, both useful and harmful.

The negative or positive charges on colloid particles make them able to hold cations or anions, respectively, to their surfaces. The charges result from four sources.

- Isomorphous substitution occurs in clay during its formation, when lower-valence cations substitute for higher-valence cations in the crystal structure. Substitutions in the outermost layers are more effective than for the innermost layers, as the charge strength drops off as the square of the distance. The net result is oxygen atoms with net negative charge and the ability to attract cations.
- Edge-of-clay oxygen atoms are not in balance ionically as the tetrahedral and octahedral structures are incomplete.
- Hydroxyls may substitute for oxygens of the silica layers. When the hydrogens of the clay hydroxyls are ionised into solution, they leave the oxygen with a negative charge.
- Hydrogens of humus hydroxyl groups may be ionised into solution, leaving an oxygen with a negative charge.

Cations held to the negatively charged colloids resist being washed downward by water and out of reach of plants' roots, thereby preserving the fertility of soils in areas of moderate rainfall and low temperatures.

There is a hierarchy in the process of cation exchange on colloids, as they differ in the strength of adsorption by the colloid and hence their ability to replace one another. If present in equal amounts in the soil water solution:

Al^{3+} replaces H^+ replaces Ca^{2+} replaces Mg^{2+} replaces K^+ same as NH^{4+} replaces Na^+

If one cation is added in large amounts, it may replace the others by the sheer force of its numbers. This is called mass action. This is largely what occurs with the addition of fertiliser.

As the soil solution becomes more acidic (low pH, and an abundance of H^+), the other cations more weakly bound to colloids are pushed into solution as hydrogen ions occupy those sites. A low pH may cause hydrogen of hydroxyl groups to be pulled into solution, leaving charged sites on the colloid available to be occupied by other cations. This ionisation of hydroxyl groups on the surface of soil colloids creates what is described as pH-dependent charges. Unlike permanent charges developed by isomorphous substitution, pH-dependent charges are variable and increase with increasing pH. Freed cations can be made available to plants but are also prone to be leached from the soil, possibly making the soil less fertile. Plants are able to excrete H^+ into the soil and by that means, change the pH of the soil near the root and push cations off the colloids, thus making those available to the plant.

Cation Exchange Capacity (CEC)

Cation exchange capacity should be thought of as the soil's ability to remove cations from the soil water solution and sequester those to be exchanged later as the plant roots release hydrogen ions to the solution. CEC is the amount of exchangeable hydrogen cation (H^+) that will combine with 100 grams dry weight of soil and whose measure is one milliequivalents per 100 grams of soil (1 meq/100 g). Hydrogen ions have a single charge and one-thousandth of a gram of hydrogen ions per 100 grams dry soil gives a measure of one milliequivalent of hydrogen ion. Calcium, with an atomic weight 40 times that of hydrogen and with a valence of two, converts to (40/2) x 1 milliequivalent = 20 milliequivalents of hydrogen ion per 100 grams of dry soil or 20 meq/100 g. The modern measure of CEC is expressed as centimoles of positive charge per kilogram (cmol/kg) of oven-dry soil.

Most of the soil's CEC occurs on clay and humus colloids, and the lack of those in hot, humid, wet climates, due to leaching and decomposition respectively, explains the relative sterility of tropical soils. Live plant roots also have some CEC.

Cation exchange capacity for soils; soil textures; soil colloids

Soil	**State**	**CEC meq/100 g**
Charlotte fine sand	Florida	1.0
Ruston fine sandy loam	Texas	1.9
Glouchester loam	New Jersey	11.9
Grundy silt loam	Illinois	26.3
Gleason clay loam	California	31.6
Susquehanna clay loam	Alabama	34.3
Davie mucky fine sand	Florida	100.8
Sands	------	1 - 5
Fine sandy loams	------	5-10
Loams and silt loams	-----	5-15
Clay loams	-----	15-30
Clays	-----	over 30
Sesquioxides	-----	0-3
Kaolinite	-----	3-15
Illite	-----	25-40
Montmorillonite	-----	60-100
Vermiculite (similar to illite)	-----	80-150
Humus	-----	100-300

Anion Exchange Capacity (AEC)

Anion exchange capacity should be thought of as the soil's ability to remove anions from the soil water solution and sequester those for later exchange as the plant roots release carbonate anions to the soil water solution. Those colloids which have low CEC tend to have some AEC. Amorphous and sesquioxide clays have the highest AEC, followed by the iron oxides. Levels of AEC are much lower than for CEC. Phosphates tend to be held at anion exchange sites.

Iron and aluminum hydroxide clays are able to exchange their hydroxide anions (OH^-) for other anions. The order reflecting the strength of anion adhesion is as follows:

$H_2PO_4^-$ replaces SO_4^{2-} replaces NO_3^- replaces Cl^-

The amount of exchangeable anions is of a magnitude of tenths to a few milliequivalents per 100 g dry soil. As pH rises, there are relatively more hydroxyls, which will displace anions from the colloids and force them into solution and out of storage; hence AEC decreases with increasing pH (alkalinity).

Soil Reaction (Ph)

Soil reactivity is expressed in terms of pH and is a measure of the acidity or alkalinity of the soil. More precisely, it is a measure of hydrogen ion concentration in an aqueous solution and ranges in values from 0 to 14 (acidic to basic) but practically speaking for soils, pH ranges from 3.5 to 9.5, as pH values beyond those extremes are toxic to life forms.

Soil pH

At 25 °C an aqueous solution that has a pH of 3.5 has $10^{-3.5}$ moles H^+ (hydrogen ions) per litre of solution (and also $10^{-10.5}$ mole/litre OH^-). A pH of 7, defined as neutral, has 10^{-7} moles hydrogen ions per litre of solution and also 10^{-7} moles of OH^- per litre; since the two concentrations are equal, they are said to neutralise each other. A pH of 9.5 has $10^{-9.5}$ moles hydrogen ions per litre of solution (and also $10^{-2.5}$ mole per litre OH^-). A pH of 3.5 has one million times more hydrogen ions per litre than a solution with pH of 9.5 (9.5 - 3.5 = 6 or 10^6) and is more acidic.

The effect of pH on a soil is to remove from the soil or to make available certain ions. Soils with high acidity tend to have toxic amounts of aluminium and manganese. Plants which need calcium need moderate alkalinity, but most minerals are more soluble in acid soils. Soil organisms are hindered by high acidity, and most agricultural crops do best with mineral soils of pH 6.5 and organic soils of pH 5.5.

In high rainfall areas, soils tend to acidity as the basic cations are forced off the soil colloids by the mass action of hydrogen ions from the rain as those attach to the colloids. High rainfall rates can then wash the nutrients out, leaving the soil sterile. Once the colloids are saturated with H^+, the addition of any more hydrogen ions or aluminum hydroxyl cations drives the pH even lower (more acidic) as the soil has been left with no buffering capacity. In areas of extreme rainfall and high temperatures, the clay and humus may be washed out, further reducing the buffering capacity of the soil. In low rainfall areas, unleached calcium pushes pH to 8.5 and with the addition of exchangeable sodium, soils may reach pH 10. Beyond a pH of 9, plant growth is reduced. High pH results in low micro-nutrient mobility, but water-soluble chelates of those nutrients can correct the deficit. Sodium can be reduced by the addition of gypsum (calcium sulphate) as calcium adheres to clay more tightly than does sodium causing sodium to be pushed into the soil water solution where it can be washed out by an abundance of water.

Base Saturation Percentage

There are acid-forming cations (hydrogen and aluminium) and there are base-forming cations. The fraction of the base-forming cations that occupy positions on the soil colloids is called the base saturation percentage. If a soil has a CEC of 20 meq and 5 meq are aluminium and hydrogen cations (acid-forming), the remainder of positions on the colloids (20-5 = 15 meq) are assumed occupied by base-forming cations, so that the percentage base saturation is 15/20 x 100% = 75% (the compliment 25% is assumed acid-forming cations). When the soil pH is 7 (neutral), base saturation is 100 percent and there are no hydrogen ions stored on the colloids. Base saturation is almost in direct proportion to pH (increases with increasing pH). It is of use in calculating the amount of lime needed to neutralise an acid soil. The amount of lime needed to neutralize a soil must take account of the amount of acid forming ions on the colloids not just those in the soil water solution. The addition of enough lime to neutralize the soil water solution will be insufficient to change the pH, as the acid forming cations stored on the soil colloids will tend to restore the original pH condition as they are pushed off those colloids by the calcium of the added lime.

Buffering of Soils

The resistance of soil to changes in pH, as a result of the addition of acid or basic material, is a measure of the buffering capacity of a soil and increases as the CEC increases. Hence, pure sand has almost no buffering ability, while soils high in colloids have high buffering capacity. Buffering occurs by cation exchange and neutralisation.

The addition of a small amount highly basic aqueous ammonia to a soil will cause the ammonium to displace hydrogen ions from the colloids, and the end product is water and colloidally fixed ammonium, but no permanent change overall in soil pH.

The addition of a small amount of lime, $CaCO_3$, will displace hydrogen ions from the soil colloids, causing the fixation of calcium to colloids and the evolution of $CO_2$and water, with no permanent change in soil pH.

The addition of carbonic acid (the solution of CO_2 in water) will displace calcium from colloids, as hydrogen ions are fixed to the colloids, evolving water and slightly alkaline (temporary increase in pH) highly soluble calcium bicarbonate, which will then precipitate as lime ($CaCO_3$) and water at a lower level in the soil profile, with the result of no permanent change in soil pH.

All of the above are examples of the buffering of soil pH. The general principal is that an increase in a particular cation in the soil water solution will cause that cation to be fixed to colloids (buffered) and a decrease in solution of that cation will cause it to be withdrawn from the colloid and moved into solution (buffered). The degree of buffering is limited by the CEC of the soil; the greater the CEC, the greater the buffering capacity of the soil.

Nutrients

Sixteen elements or nutrients are essential for plant growth and reproduction. They are carbon C, hydrogen H, oxygen O, nitrogen N, phosphorus P, potassium K, sulfur S, calcium Ca, magnesium Mg, iron Fe, boron B, manganese Mn, copper Cu, zinc Zn, molybdenum Mo, and chlorine Cl. Nutrients required for plants to complete their life cycle are considered essential nutrients. Nutrients that enhance the growth of plants but are not necessary to complete the plant's life cycle are considered non-essential. With the exception of carbon, hydrogen and oxygen, which are supplied by carbon dioxide and water, the nutrients derive originally from the mineral component of the soil.

Plant uptake of nutrients can only proceed when they are present in a plant-available form. In most situations, nutrients are absorbed in an ionic form from (or together with) soil water. Although minerals are the origin of most nutrients, and the bulk of most nutrient elements in the soil is held in crystalline form within primary and secondary minerals, they weather too slowly to support rapid plant growth. For example, The application of finely ground minerals, feldspar and apatite, to soil seldom provides the necessary amounts of potassium and phosphorus at a rate sufficient for good plant growth, as most of the nutrients remain bound in the crystals of those minerals.

The nutrients adsorbed onto the surfaces of clay colloids and soil organic matter provide a more accessible reservoir of many plant nutrients (e.g. K, Ca, Mg, P, Zn). As plants absorb the nutrients from the soil water, the soluble pool is replenished from the surface-bound pool. The decomposition of soil organic matter by microorganisms is another mechanism whereby the soluble pool of nutrients is replenished - this is important for the supply of plant-available N, S, P, and B from soil.

Gram for gram, the capacity of humus to hold nutrients and water is far greater than that of clay minerals. All in all, small amounts of humus may remarkably increase the soil's capacity to promote plant growth.

Plant nutrients, their chemical symbols, and the ionic forms common in soils and available for plant uptake

Element	**Symbol**	**Ion or molecule**
Carbon	C	CO_2 (mostly through leaves)
Hydrogen	H	H^+, HOH (water)
Oxygen	O	O^{2-}, OH^-, CO_3^{2-}, SO_4^{2-}, CO_2
Phosphorus	P	$H_2PO_4^-$, HPO_4^{2-} (phosphates)
Potassium	K	K^+
Nitrogen	N	NH_4^+, NO_3^- (ammonium, nitrate)
Sulfur	S	SO_4^{2-}
Calcium	Ca	Ca^{2+}
Iron	Fe	Fe^{2+}, Fe^{3+} (ferrous, ferric)
Magnesium	Mg	Mg^{2+}

Element	Symbol	Ion or molecule
Boron	B	H_3BO_3, $H_2BO_3^-$, $B(OH)_4^-$
Manganese	Mn	Mn^{2+}
Copper	Cu	Cu^{2+}
Zinc	Zn	Zn^{2+}
Molybdenum	Mo	MoO_4^{2-} (molybdate)
Chlorine	Cl	Cl^- (chloride)

Soil Processes Important for Nutrient Uptake

Nutrients in the soil are taken up by the plant through its roots. To be taken up by a plant, a nutrient element must be located near the root surface; however, the supply of nutrients in contact with the root is rapidly depleted. There are three basic mechanisms whereby nutrient ions dissolved in the soil solution are brought into contact with plant roots:

- Mass flow of water
- Diffusion within water
- Interception by root growth

All three mechanisms operate simultaneously, but one mechanism or another may be most important for a particular nutrient. For example, in the case of calcium, which is generally plentiful in the soil solution, mass flow alone can usually bring sufficient amounts to the root surface. However, in the case of phosphorus, diffusion is needed to supplement mass flow. For the most part, nutrient ions must travel some distance in the soil solution to reach the root surface. This movement can take place by mass flow, as when dissolved nutrients are carried along with the soil water flowing toward a root that is actively drawing water from the soil. In this type of movement, the nutrient ions are somewhat analogous to leaves floating down a stream. In addition, nutrient ions continually move by diffusion from areas of greater concentration toward the nutrient-depleted areas of lower concentration around the root surface. That process is due to random motion of molecules. By this means, plants can continue to take up nutrients even at night, when water is only slowly absorbed into the roots as transpiration has almost stopped. Finally, root interception comes into play as roots continually grow into new, undepleted soil.

Estimated relative importance of mass flow, diffusion and root interception as mechanisms in supplying plant nutrients to corn plant roots in soils

Nutrient	Approximate percentage supplied by:		
	Mass flow	Root interception	Diffusion
Nitrogen	98.8	1.2	0
Phosphorus	6.3	2.8	90.9
Potassium	20.0	2.3	77.7
Calcium	71.4	28.6	0
Sulfur	95.0	5.0	0
Molybdenum	95.2	4.8	0

In the above table, phosphorus and potassium nutrients move more by diffusion than they do by mass flow in the soil water solution, as they are rapidly taken up by the roots creating a concentration of almost zero near the roots (the plants cannot transpire enough water to draw more of those nutrients near the roots). The very steep concentration gradient is of greater influence in the movement of those ions than is the movement of those by mass flow. The movement by mass flow requires the transpiration of water from the plant causing water and solution ions to also move toward the roots. Movement by root interception is slowest as the plants must extend their roots.

Plants move ions out of their roots in an effort to move nutrients in from the soil. Hydrogen H^+ is exchanged for other cations, and carbonate (HCO_3^-) and hydroxide (OH^-) anions are exchanged for nutrient anions. As plant roots remove nutrients from the soil water solution, they are replenished as other ions move off of clay and humus (by ion exchange or desorption), are added from the weathering of soil minerals, and are released by the decomposition of soil organic matter. Plants derive a large proportion of their anion nutrients from decomposing organic matter, which typically holds about 95 percent of the soil nitrogen, 5 to 60 percent of the soil phosphorus and about 80 percent of the soil sulfur. Where crops are produced, the replenishment of nutrients in the soil must usually be augmented by the addition of fertilizer or organic matter.

Because nutrient uptake is an active metabolic process, conditions that inhibit root metabolism may also inhibit nutrient uptake. Examples of such conditions include waterlogging or soil compaction resulting in poor soil aeration, excessively high or low soil temperatures, and above-ground conditions that result in low translocation of sugars to plant roots.

Carbon

Measuring soil respiration in the field using an SRS2000 system.

Plants obtain their carbon from atmospheric carbon dioxide. About 45% of a plant's dry mass is carbon; plant residues typically have a carbon to nitrogen ratio (C/N) of between 13:1 and 100:1. As the soil organic material is digested by arthropods and micro-organisms, the C/N decreases as the carbonaceous material is metabolized and carbon dioxide (CO_2) is released as a byproduct which then finds its way out of the soil and into the atmosphere. The nitrogen is sequestered in the bodies of the living matter of those decomposing organisms and so it builds up in the soil. Normal CO_2 concentration in the atmosphere is 0.03%, this can be the factor limiting plant growth. In a field of maize on a still day during high light conditions in the growing season, the CO_2 concentration drops very low, but under such conditions the crop could use up to 20 times the normal concentration. The respiration of CO_2 by soil micro-organisms decomposing soil organic matter contributes an important amount of CO_2 to the photosynthesising plants. Within the soil, CO_2 concentration is 10 to 100 times that of atmospheric levels but may rise to toxic levels if the soil porosity is low or if diffusion is impeded by flooding.

Nitrogen

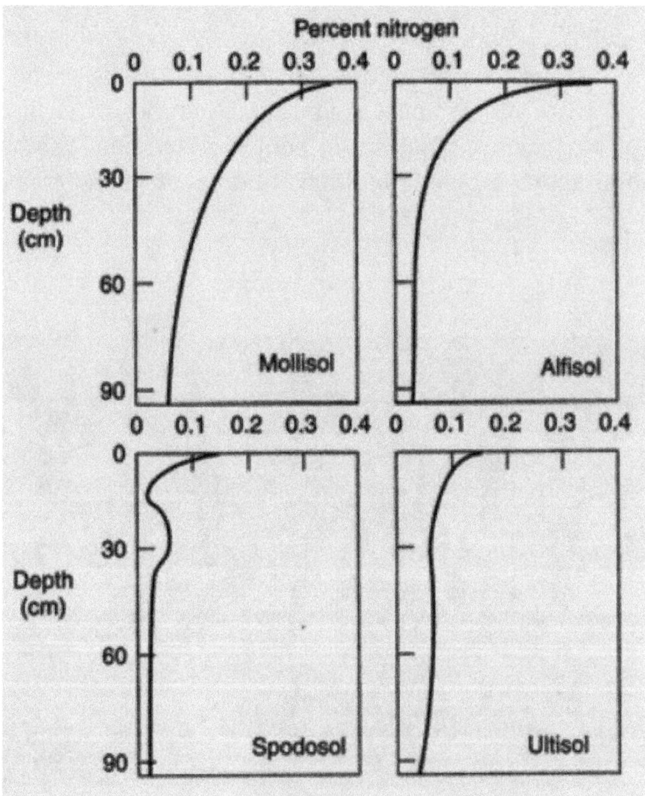

Generalization of percent soil nitrogen by soil order

Nitrogen is the most critical element obtained by plants from the soil and nitrogen deficiency often limits plant growth. Plants can use the nitrogen as either the ammonium cation (NH_4^+) or the anion nitrate (NO_3^-). Usually, most of the nitrogen in soil is bound within organic compounds that make up the soil organic matter, and must be mineralized to the ammonium or nitrate form before it can be taken up by most plants. The total nitrogen content depends largely on the soil organic matter content, which in turn depends on the climate, vegetation, topography, age and soil management. Soil nitrogen typically decreases by 0.2 to 0.3% for every temperature increase by 10 °C. Usually, grassland soils contain more soil nitrogen than forest soils. Cultivation decreases soil nitrogen by exposing soil organic matter to decomposition by microorganisms, and soils under no-tillage maintain more soil nitrogen than tilled soils.

Some micro-organisms are able to metabolise organic matter and release ammonium in a process called *mineralisation*. Others take free ammonium and oxidise it to nitrate. Nitrogen-fixing bacteria are capable of metabolising N_2 into the form of ammonia in a process called nitrogen fixation. Both ammonium and nitrate can be *immobilized* by their incorporation into the microbes' living cells, where it is temporarily sequestered in the form of amino acids and protein. Nitrate may also be lost from the soil when bacteria metabolise it to the gases N_2 and N_2O. The loss of gaseous forms of nitrogen to the atmosphere due to microbial action is called *denitrification*. Nitrogen may also be *leached* from the soil if it is in the form of nitrate or lost to the atmosphere as ammonia due to a chemical reaction of ammonium with alkaline soil by way of a process called *volatilisation*. Ammonium may also be sequestered in clay by *fixation*. A small amount of nitrogen is added to soil by rainfall.

Nitrogen Gains

In the process of mineralisation, microbes feed on organic matter, releasing ammonia (NH_3), ammonium (NH_4^+) and other nutrients. As long as the carbon to nitrogen ratio (C/N) of fresh residues in the soil is above 30:1, nitrogen will be in short supply and other bacteria will feed on the ammonium and incorporate its nitrogen into their cells in the immobilization process. In that form the nitrogen is said to be *immobilised*. Later, when such bacteria die, they too are *mineralised* and some of the nitrogen is released as ammonium and nitrate. If the C/N is less than 15, ammonia is freed to the soil, where it may be used by bacteria which oxidise it to nitrate (nitrification). Bacteria may on average add 25 pounds (11 kg) nitrogen per acre, and in an unfertilised field, this is the most important source of usable nitrogen. In a soil with 5% organic matter perhaps 2 to 5% of that is released to the soil by such decomposition. It occurs fastest in warm, moist, well aerated soil. The mineralisation of 3% of the organic material of a soil that is 4% organic matter overall, would release 120 pounds (54 kg) of nitrogen as ammonium per acre.

Carbon/Nitrogen Ratio of Various Organic Materials

Organic Material	C:N Ratio
Alfalfa	13
Bacteria	4
Clover, green sweet	16
Clover, mature sweet	23
Fungi	9

Organic Material	C:N Ratio
Forest litter	30
Humus in warm cultivated soils	11
Legume-grass hay	25
Legumes (alfalfa or clover), mature	20
Manure, cow	18
Manure, horse	16–45
Manure, human	10
Oat straw	80
Straw, cornstalks	90
Sawdust	250

In nitrogen fixation, rhizobium bacteria convert N_2 to ammonia (NH_3). Rhizobia share a symbiotic relationship with host plants, since rhizobia supply the host with nitrogen and the host provides rhizobia with nutrients and a safe environment. It is estimated that such symbiotic bacteria in the root nodules of legumes add 45 to 250 pounds of nitrogen per acre per year, which may be sufficient for the crop. Other, free-living nitrogen-fixing bacteria and blue-green algae live independently in the soil and release nitrate when their dead bodies are converted by way of mineralisation.

Some amount of usable nitrogen is fixed by lightning as nitric oxide (NO) and nitrogen dioxide (NO_2^-). Nitrogen dioxide is soluble in water to form nitric acid (HNO_3) solution of H^+ and NO_3^-. Ammonia, NH_3, previously released from the soil or from combustion, may fall with precipitation as nitric acid at a rate of about five pounds nitrogen per acre per year.

Nitrogen Sequestration

When bacteria feed on soluble forms of nitrogen (ammonium and nitrate), they temporarily sequester that nitrogen in their bodies in a process called *immobilisation*. At a later time when those bacteria die, their nitrogen may be released as ammonium by the processes of mineralisation.

Protein material is easily broken down, but the rate of its decomposition is slowed by its attachment to the crystalline structure of clay and when trapped between the clay layers. The layers are small enough that bacteria cannot enter. Some organisms can exude extracellular enzymes that can act on the sequestered proteins. However, those enzymes too may be trapped on the clay crystals.

Ammonium fixation occurs when ammonium pushes potassium ions from between the layers of clay such as illite or montmorillonite. Only a small fraction of soil nitrogen is held this way.

Nitrogen Losses

Usable nitrogen may be lost from soils when it is in the form of nitrate, as it is easily leached. Further losses of nitrogen occur by denitrification, the process whereby soil bacteria convert nitrate (NO_3^-) to nitrogen gas, N_2 or N_2O. This occurs when poor soil aeration limits free oxygen, forcing bacteria to use the oxygen in nitrate for their respiratory process. Denitrification increases when oxidisable organic material is available and when soils are warm and slightly acidic. Denitrification may vary throughout a soil as the aeration varies from place to place. Denitrification may cause the loss of 10 to 20 percent of the available nitrates within a day and when conditions are favourable to that process, losses of up to 60 percent of nitrate applied as fertiliser may occur.

Ammonium volatilisation occurs when ammonium reacts chemically with an alkaline soil, converting NH_4^+ to NH_3. The application of ammonium fertiliser to such a field can result in volatilisation losses of as much as 30 percent.

Phosphorus

After nitrogen, phosphorus is probably the element most likely to be deficient in soils. The soil mineral apatite is the most common mineral source of phosphorus. While there is on average 1000 lb of phosphorus per acre in the soil, it is generally unavailable in the form of phosphates of low solubility. Total phosphorus is about 0.1 percent by weight of the soil, but only one percent of that is available. Of the part available, more than half comes from the mineralisation of organic matter. Agricultural fields may need to be fertilised to make up for the phosphorus that has been removed in the crop.

When phosphorus does form solubilised ions of $H_2PO_4^-$, they rapidly form insoluble phosphates of calcium or hydrous oxides of iron and aluminum. Phosphorus is largely immobile in the soil and is not leached but actually builds up in the surface layer if not cropped. The application of soluble fertilisers to soils may result in zinc deficiencies as zinc phosphates form. Conversely, the application of zinc to soils may immobilise phosphorus again as zinc phosphate. Lack of phosphorus may interfere with the normal opening of the plant leaf stomata, resulting in plant temperatures 10 percent higher than normal. Phosphorus is most available when soil pH is 6.5 in mineral soils and 5.5 in organic soils.

Potassium

The amount of potassium in a soil may be as much as 80,000 lb per acre-foot, of which only 150 lb is available for plant growth. Common mineral sources of potassium are the mica biotite and potassium feldspar, $KAlSi_3O_8$. When solubilised, half will be held as exchangeable cations on clay while the other half is in the soil water solution. Potassium fixation often occurs when soils dry and the potassium is bonded between layers of illite clay. Under certain conditions, dependent on the soil texture, intensity of drying, and initial amount of exchangeable potassium, the fixed percentage may be as much as 90 percent within ten minutes. Potassium may be leached from soils low in clay.

Calcium

Calcium is one percent by weight of soils and is generally available but may be low as it is soluble and can be leached. It is thus low in sandy and heavily leached soil or strongly acidic mineral soil. Calcium is supplied to the plant in the form of exchangeable ions and moderately soluble minerals. Calcium is more available on the soil colloids than is potassium because the common mineral calcite, $CaCO_3$, is more soluble than potassium-bearing minerals.

Magnesium

Magnesium is one of the dominant exchangeable cations in most soils (as are calcium and potassium). Primary minerals that weather to release magnesium include hornblende, biotite and vermiculite. Soil magnesium concentrations are generally sufficient for optimal plant growth, but highly weathered and sandy soils may be magnesium deficient due to leaching by heavy precipitation.

Sulfur

Most sulfur is made available to plants, like phosphorus, by its release from decomposing organic matter. Deficiencies may exist in some soils (especially sandy soils) and if cropped, sulfur needs to be added. The application of large quantities of nitrogen to fields that have marginal amounts of sulfur may cause sulfur deficiency in the rapidly growing plants by the plant's growth outpacing the supply of sulfur. A 15-ton crop of onions uses up to 19 lb of sulfur and 4 tons of alfalfa uses 15 lb per acre. Sulfur abundance varies with depth. In a sample of soils in Ohio, United States, the sulfur abundance varied with depths, 0-6 inches, 6-12 inches, 12-18 inches, 18-24 inches in the amounts: 1056, 830, 686, 528 lb per acre respectively.

Micronutrients

The micronutrients essential for plant life, in their order of importance, include iron, manganese, zinc, copper, boron, chlorine and molybdenum. The term refers to plants' needs, not to their abundance in soil. They are required in very small amounts but are essential to plant health in that most are required parts of some enzyme system which speeds up plants' metabolisms. They are generally available in the mineral component of the soil, but the heavy application of phosphates can cause a deficiency in zinc and iron by the formation of insoluble zinc and iron phosphates. Iron deficiency may also result from excessive amounts of heavy metals or calcium minerals (lime) in the soil. Excess amounts of soluble boron, molybdenum and chloride are toxic.

Non-Essential Nutrients

Nutrients which enhance the health but whose deficiency does not stop the life cycle of plants include: cobalt, strontium, vanadium, silicon and nickel.As their importance are evaluated they may be added to the list of essential plant nutrients.

Soil Organic Matter

Soil organic matter is made up of organic compounds and includes plant, animal and microbial material, both living and dead. A typical soil has a biomass composition of 70% microorganisms, 22% macrofauna, and 8% roots. The living component of an acre of soil may include 900 lb of earthworms, 2400 lb of fungi, 1500 lb of bacteria, 133 lb of protozoa and 890 lb of arthropods and algae.

A small part of the organic matter consists of the living cells such as bacteria, molds, and actinomycetes that work to break down the dead organic matter. Were it not for the action of these micro-organisms, the entire carbon dioxide part of the atmosphere would be sequestered as organic matter in the soil.

Chemically, organic matter is classed as follows:

- Polysaccharides
    - cellulose
    - hemicellulose
    - starch
    - pectin
- Lignins
- Proteins

Most living things in soils, including plants, insects, bacteria, and fungi, are dependent on organic matter for nutrients and/or energy. Soils have organic compounds in varying degrees of decomposition which rate is dependent on the temperature, soil moisture, and aeration. Bacteria and fungi feed on the raw organic matter, which are fed upon by amoebas, which in turn are fed upon by nematodes and arthropods. Organic matter holds soils open, allowing the infiltration of air and water, and may hold as much as twice its weight in water. Many soils, including desert and rocky-gravel soils, have little or no organic matter. Soils that are all organic matter, such as peat (histosols), are infertile. In its earliest stage of decomposition, the original organic material is often called raw organic matter. The final stage of decomposition is called humus.

In grassland, much of the organic matter added to the soil is from the deep, fibrous, grass root systems. By contrast, tree leaves falling on the forest floor are the principal source of soil organic matter in the forest. Another difference is the frequent occurrence in the grasslands of fires that destroy large amounts of aboveground material but stimulate even greater contributions from roots. Also, the much greater acidity under any forests inhibits the action of certain soil organisms that otherwise would mix much of the surface litter into the mineral soil. As a result, the soils under grasslands generally develop a thicker A horizon with a deeper distribution of organic matter than in comparable soils under forests, which characteristically store most of their organic matter in the forest floor (O horizon) and thin A horizon.

Humus

Humus refers to organic matter that has been decomposed by soil flora and fauna to the point where it is resistant to further breakdown. Humus usually constitutes only five percent of the soil or less by volume, but it is an essential source of nutrients and adds important textural qualities crucial to soil health and plant growth. Humus also hold bits of undecomposed organic matter which feed arthropods and worms which further improve the soil. The end product, humus, is soluble in water and forms a weak acid that can attack silicate minerals. Humus is a colloid with a high cation and anion exchange capacity that on a dry weight basis is many times greater than that of clay colloids. It also acts as a buffer, like clay, against changes in pH and soil moisture.

Humic acids and fulvic acids, which begin as raw organic matter, are important constituents of humus. After the death of plants and animals, microbes begin to feed on the residues, resulting finally in the formation of humus. With decomposition, there is a reduction of water-soluble constituents, cellulose and hemicellulose, and nutrients such as nitrogen, phosphorus, and sulfur. As the residues break down, only stable molecules made of aromatic carbon rings, oxygen and hydrogen remain in the form of humin, lignin and lignin complexes collectively called humus. While the structure of humus has few nutrients, it is able to attract and hold cation and anion nutrients by weak bonds that can be released into the soil solution in response to changes in soil pH.

Lignin is resistant to breakdown and accumulates within the soil. It also reacts with amino acids, which further increases its resistance to decomposition, including enzymatic decomposition by microbes. Fats and waxes from plant matter have some resistance to decomposition and persist in soils for a while. Clay soils often have higher organic contents that persist longer than soils without clay as the organic molecules adhere to and are stabilised by the clay. Proteins normally decompose readily, but when bound to clay particles, they become more resistant to decomposition. Clay particles also absorb the enzymes exuded by microbes which would normally break down proteins. The addition of organic matter to clay soils can render that organic matter and any added nutrients inaccessible to plants and microbes for many years. High soil tannin (polyphenol) content can cause nitrogen to be sequestered in proteins or cause nitrogen immobilisation.

Humus formation is a process dependent on the amount of plant material added each year and the type of base soil. Both are affected by climate and the type of organisms present. Soils with humus can vary in nitrogen content but typically have 3 to 6 percent nitrogen. Raw organic matter, as a reserve of nitrogen and phosphorus, is a vital component affecting soil fertility. Humus also absorbs water, and expands and shrinks between dry and wet states, increasing soil porosity. Humus is less stable than the soil's mineral constituents, as it is reduced by microbial decomposition, and over time its concentration diminshes without the addition of new organic matter. However, humus may persist over centuries if not millennia.

Climate and Organic Matter

The production, accumulation and degradation of organic matter are greatly dependent on climate. Temperature, soil moisture and topography are the major factors affecting the accumulation of organic matter in soils. Organic matter tends to accumulate under wet or cold conditions where decomposer activity is impeded by low temperature or excess moisture which results in anaerobic conditions. Conversely, excessive rain and high temperatures of tropical climates enables rapid decomposition of organic matter and leaching of plant nutrients; forest ecosystems on these soils rely on efficient recycling of nutrients and plant matter to maintain their productivity. Excessive slope may encourage the erosion of the top layer of soil which holds most of the raw organic material that would otherwise eventually become humus.

Plant Residue in Soil

Typical types and percentages of plant residue components

- Cellulose (45%)
- Lignin (20%)
- Hemicellulose (18%)
- Protein (8%)
- Sugars and starches (5%)
- Fats and waxes (2%)

Cellulose and hemicellulose undergo fast decomposition by fungi and bacteria, with a half-life of 12–18 days in a temperate climate. Brown rot fungi can decompose the cellulose and hemicellulose, leaving the lignin and phenolic compounds behind. Starch, which is an energy storage system for plants, undergoes fast decomposition by bacteria and fungi. Lignin consists of polymers composed of 500 to 600 units with a highly branched, amorphous structure. Lignin undergoes very slow decomposition, mainly by white rot fungi and actinomycetes; its half-life under temperate conditions is about six months.

Soil Horizons

A horizontal layer of the soil, whose physical features, composition and age are distinct from those above and beneath, are referred to as a soil horizon. The naming of a horizon is based on the type of material of which it is composed. Those materials reflect the duration of specific processes of soil formation. They are labelled using a shorthand notation of letters and numbers which describe the horizon in terms of its colour, size, texture, structure, consistency, root quantity, pH, voids, boundary characteristics and presence of nodules or concretions. Few soil profiles have all the major horizons. Some may have only one horizon.

The exposure of parent material to favourable conditions produces mineral soils that are marginally suitable for plant growth. That growth often results in the accumulation of organic residues. The accumulated organic layer called the O horizon produces a more active soil due to the effect of the organisms that live within it. Organisms colonise and break down organic materials, making available nutrients upon which other plants and animals can live. After sufficient time, humus moves downward and is deposited in a distinctive organic surface layer called the A horizon.

Classification

Soil is classified into categories in order to understand relationships between different soils and to determine the suitability of a soil for a particular use. One of the first classification systems was developed by the Russian scientist Dokuchaev around 1880. It was modified a number of times by American and European researchers, and developed into the system commonly used until the 1960s. It was based on the idea that soils have a particular morphology based on the materials and factors that form them. In the 1960s, a different classification system began to emerge which focused on soil morphology instead of parental materials and soil-forming factors. Since then it has undergone further modifications. The World Reference Base for Soil Resources (WRB) aims to establish an international reference base for soil classification.

Soil Classification Systems

Australia

There are fourteen soil orders at the top level of the Australian Soil Classification. They are: Anthroposols, Organosols, Podosols, Vertosols, Hydrosols, Kurosols, Sodosols, Chromosols, Calcarosols, Ferrosols, Dermosols, Kandosols, Rudosols and Tenosols.

European Union

The EU's soil taxonomy is based on a new standard soil classification in the World Reference Base for Soil Resources produced by the UN's Food and Agriculture Organization. According to this, the major soils in the European Union are:

- Acrisols
- Albeluvisols
- Andosols
- Anthrosols
- Arenosols
- Calcisols
- Cambisols
- Chernozems
- Fluvisols
- Gleysols

- Gypsisols
- Histosols
- Kastanozems
- Leptosols
- Luvisols
- Phaeozems
- Planosols
- Podzols
- Regosols
- Solonchaks
- Solonetzes
- Umbrisols
- Vertisols

USA

A taxonomy is an arrangement in a systematic manner; the USDA soil taxonomy has six levels of classification. They are, from most general to specific: order, suborder, great group, subgroup, family and series. Soil properties that can be measured quantitatively are used in this classification system — they include: depth, moisture, temperature, texture, structure, cation exchange capacity, base saturation, clay mineralogy, organic matter content and salt content. There are 12 soil orders (the top hierarchical level) in the USDA soil taxonomy. The names of the orders end with the suffix *-sol*. The criteria for the different soil orders include properties that reflect major differences in the genesis of soils. The orders are:

- Alfisol - soils with aluminium and iron. They have horizons of clay accumulation, and form where there is enough moisture and warmth for at least three months of plant growth. They constitute 10.1% of soils worldwide.
- Andisol - volcanic ash soils. They are young and very fertile. They cover 1% of the world's ice-free surface.
- Aridisol - dry soils forming under desert conditions which have fewer than 90 consecutive days of moisture during the growing season and are nonleached. They include nearly 12% of soils on Earth. Soil formation is slow, and accumulated organic matter is scarce. They may have subsurface zones of caliche or duripan. Many aridisols have well-developed Bt horizons showing clay movement from past periods of greater moisture.
- Entisol - recently formed soils that lack well-developed horizons. Commonly found on unconsolidated river and beach sediments of sand and clay or volcanic ash, some have an A horizon on top of bedrock. They are 18% of soils worldwide.
- Gelisol - permafrost soils with permafrost within two metres of the surface or gelic materials and permafrost within one metre. They constitute 9.1% of soils

worldwide.

- Histosol - organic soils, formerly called bog soils, are 1.2% of soils worldwide.
- Inceptisol - young soils. They have subsurface horizon formation but show little eluviation and illuviation. They constitute 15% of soils worldwide.
- Mollisol - soft, deep, dark fertile soil formed in grasslands and some hardwood forests with very thick A horizons. They are 7% of soils worldwide.
- Oxisol - are heavily weathered, are rich in iron and aluminum oxides (sesquioxides) or kaolin but low in silica. They have only trace nutrients due to heavy tropical rainfall and high temperatures and low CEC of the remaining clays. They are 7.5% of soils worldwide.
- Spodosol - acid soils with organic colloid layer complexed with iron and aluminium leached from a layer above. They are typical soils of coniferous and deciduous forests in cooler climates. They constitute 4% of soils worldwide.
- Ultisol - acid soils in the humid tropics and subtropics, which are depleted in calcium, magnesium and potassium (important plant nutrients). They are highly weathered, but not as weathered as Oxisols. They make up 8% of the soil worldwide.
- Vertisol - inverted soils. They are clay-rich and tend to swell when wet and shrink upon drying, often forming deep cracks into which surface layers can fall. They are difficult to farm or to construct roads and buildings due to their high expansion rate. They constitute 2.4% of soils worldwide.

Alfisol

Andisol

Aridisol

Entisol

Gelisol

Histisol

Inceptisol

Mollisol

Oxisol

Spodosol

Utisol

Vertisol

The percentages listed above are for land area free of ice. "Soils of Mountains", which constitute the balance (11.6%), have a mixture of those listed above, or are classified as "Rugged Mountains" which have no soil.

The above soil orders in sequence of increasing degree of development are Entisols, Inceptisols, Aridisols, Mollisols, Alfisols, Spodosols, Ultisols, and Oxisols. Histosols and Vertisols may appear in any of the above at any time during their development.

The soil suborders within an order are differentiated on the basis of soil properties and horizons which depend on soil moisture and temperature. Forty-seven suborders are recognized in the United States.

The soil great group category is a subdivision of a suborder in which the kind and sequence of soil horizons distinguish one soil from another. About 185 great groups are recognized in the United States. Horizons marked by clay, iron, humus and hard pans and soil features such as the expansion-contraction of clays (that produce self-mixing provided by clay), temperature, and marked quantities of various salts are used as distinguishing features.

The great group categories are divided into three kinds of soil subgroups: typic, intergrade and extragrade. A typic subgroup represents the basic or 'typical' concept of the great group to which the described subgroup belongs. An intergrade subgroup describes the properties that suggest how it grades towards (is similar to) soils of other soil great groups, suborders or orders. These properties are not developed or expressed well enough to cause the soil to be included within the great group towards which they grade, but suggest similarities. Extragrade features are aberrant properties which prevent that soil from being included in another soil classification. About 1,000 soil subgroups are defined in the United States.

A soil family category is a group of soils within a subgroup and describes the physical and chemical properties which affect the response of soil to agricultural management and engineering applications. The principal characteristics used to differentiate soil families include texture, mineralogy, pH, permeability, structure, consistency, the locale's precipitation pattern, and soil temperature. For some soils the criteria also specify the percentage of silt, sand and coarse fragments such as gravel, cobbles and rocks. About 4,500 soil families are recognised in the United States.

A family may contain several soil series which describe the physical location using the name of a prominent physical feature such as a river or town near where the soil sample was taken. An example would be Merrimac for the Merrimack River in New Hampshire, USA. More than 14,000 soil series are recognised in the United States. This permits very specific descriptions of soils.

A soil phase of series, originally called 'soil type' describes the soil surface texture, slope, stoniness, saltiness, erosion, and other conditions.

Uses

Soil is used in agriculture, where it serves as the anchor and primary nutrient base for plants; however, as demonstrated by hydroponics, it is not essential to plant growth if the soil-contained nutrients can be dissolved in a solution. The types of soil and available moisture determine the species of plants that can be cultivated.

Soil material is also a critical component in the mining, construction and landscape development industries. Soil serves as a foundation for most construction projects. The movement of massive volumes of soil can be involved in surface mining, road building and dam construction. Earth sheltering is the architectural practice of using soil for external thermal mass against building walls. Many building materials are soil based.

Soil resources are critical to the environment, as well as to food and fibre production. Soil provides minerals and water to plants. Soil absorbs rainwater and releases it later, thus preventing floods and drought. Soil cleans water as it percolates through it. Soil is the habitat for many organisms: the major part of known and unknown biodiversity is in the soil, in the form of invertebrates (earthworms, woodlice, millipedes, centipedes, snails, slugs, mites, springtails, enchytraeids, nematodes, protists), bacteria, archaea, fungi and algae; and most organisms living above ground have part of them (plants) or spend part of their life cycle (insects) below-ground. Above-ground and below-ground biodiversities are tightly interconnected, making soil protection of paramount importance for any restoration or conservation plan.

The biological component of soil is an extremely important carbon sink since about 57% of the biotic content is carbon. Even on desert crusts, cyanobacteria, lichens and mosses capture and sequester a significant amount of carbon by photosynthesis. Poor farming and grazing methods have degraded soils and released much of this sequestered carbon to the atmosphere. Restoring the world's soils could offset some of the huge increase in greenhouse gases causing global warming, while improving crop yields and reducing water needs.

Waste management often has a soil component. Septic drain fields treat septic tank effluent using aerobic soil processes. Landfills use soil for daily cover. Land application of waste water relies on soil biology to aerobically treat BOD.

Organic soils, especially peat, serve as a significant fuel resource; but wide areas of peat production, such as sphagnum bogs, are now protected because of patrimonial interest.

Geophagy is the practice of eating soil-like substances. Both animals and human cultures occasionally consume soil for medicinal, recreational, or religious purposes. It has been shown that some monkeys consume soil, together with their preferred food (tree foliage and fruits), in order to alleviate tannin toxicity.

Soils filter and purify water and affect its chemistry. Rain water and pooled water from ponds, lakes and rivers percolate through the soil horizons and the upper rock strata, thus becoming groundwater. Pests (viruses) and pollutants, such as persistent organic pollutants (chlorinated pesticides, polychlorinated biphenyls), oils (hydrocarbons), heavy metals (lead, zinc, cadmium), and excess nutrients (nitrates, sulfates, phosphates) are filtered out by the soil. Soil organisms metabolise them or immobilise them in their biomass and necromass, thereby incorporating them into stable humus. The physical integrity of soil is also a prerequisite for avoiding landslides in rugged landscapes.

Degradation

Land degradation refers to a human-induced or natural process which impairs the capacity of land to function. Soils degradation involves the acidification, contamination, desertification, erosion or salination.

Soil acidification is beneficial in the case of alkaline soils, but it degrades land when it lowers crop productivity and increases soil vulnerability to contamination and erosion. Soils are often initially acid because their parent materials were acid and initially low in the basic cations (calcium, magnesium, potassium and sodium). Acidification occurs when these elements are leached from the soil profile by rainfall or by the harvesting of forest or agricultural crops. Soil acidification is accelerated by the use of acid-forming nitrogenous fertilizers and by the effects of acid precipitation.

Soil contamination at low levels is often within a soil's capacity to treat and assimilate waste material. Soil biota can treat waste by transforming it; soil colloids can adsorb the waste material. Many waste treatment processes rely on this treatment capacity. Exceeding treatment capacity can damage soil biota and limit soil function. Derelict soils occur where industrial contamination or other development activity damages the soil to such a degree that the land cannot be used safely or productively. Remediation of derelict soil uses principles of geology, physics, chemistry and biology to degrade, attenuate, isolate or remove soil contaminants to restore soil functions and values. Techniques include leaching, air sparging, chemical amendments, phytoremediation, bioremediation and natural degradation.

Desertification is an environmental process of ecosystem degradation in arid and semi-arid regions, often caused by human activity. It is a common misconception that droughts cause desertification. Droughts are common in arid and semiarid lands. Well-managed lands can recover from drought when the rains return. Soil management tools include maintaining soil nutrient and organic matter levels, reduced tillage and increased cover. These practices help to control erosion and maintain productivity during periods when moisture is available. Continued land abuse during droughts, however, increases land degradation. Increased population and livestock pressure on marginal lands accelerates desertification.

Erosion of soil is caused by water, wind, ice, and movement in response to gravity. More than one kind of erosion can occur simultaneously. Erosion is distinguished from weathering, since erosion also transports eroded soil away from its place of origin (soil in transit may be described as sediment). Erosion is an intrinsic natural process, but in many places it is greatly increased by human activity, especially poor land use practices. These include agricultural activities which leave the soil bare during times of heavy rain or strong winds, overgrazing, deforestation, and improper construction activity. Improved management can limit erosion. Soil conservation techniques which are employed include changes of land use (such as replacing erosion-prone crops with grass or other soil-binding plants), changes to the timing or type of agricultural operations, terrace building, use of erosion-suppressing cover materials (including cover crops and other plants), limiting disturbance during construction, and avoiding construction during erosion-prone periods.

A serious and long-running water erosion problem occurs in China, on the middle reaches of the Yellow River and the upper reaches of the Yangtze River. From the Yellow River, over 1.6 billion tons of sediment flow each year into the ocean. The sediment originates primarily from water erosion (gully erosion) in the Loess Plateau region of northwest China.

Soil piping is a particular form of soil erosion that occurs below the soil surface. It causes levee and dam failure, as well as sink hole formation. Turbulent flow removes soil starting at the mouth of the seep flow and the subsoil erosion advances up-gradient. The term sand boil is used to describe the appearance of the discharging end of an active soil pipe.

Soil salination is the accumulation of free salts to such an extent that it leads to degradation of the agricultural value of soils and vegetation. Consequences include corrosion damage, reduced plant growth, erosion due to loss of plant cover and soil structure, and water quality problems due to sedimentation. Salination occurs due to a combination of natural and human-caused processes. Arid conditions favour salt accumulation. This is especially apparent when soil parent material is saline. Irrigation of arid lands is especially problematic. All irrigation water has some level of salinity. Irrigation, especially when it involves leakage from canals and overirrigation in the field, often raises the underlying water table. Rapid salination occurs when the land surface is within the capillary fringe of saline groundwater. Soil salinity control involves watertable control and flushing with higher levels of applied water in combination with tile drainage or another form of subsurface drainage.

Desertification

Erosion control

Reclamation

Soils which contain high levels of particular clays, such as smectites, are often very fertile. For example, the smectite-rich clays of Thailand's Central Plains are among the most productive in the world.

Many farmers in tropical areas, however, struggle to retain organic matter in the soils they work. In recent years, for example, productivity has declined in the low-clay soils of northern Thailand. Farmers initially responded by adding organic matter from termite mounds, but this was unsustainable in the long-term. Scientists experimented with adding bentonite, one of the smectite family of clays, to the soil. In field trials, conducted by scientists from the International Water Management Institute in cooperation with Khon Kaen University and local farmers, this had the effect of helping retain water and nutrients. Supplementing the farmer's usual practice with a single application of 200 kg bentonite per rai (6.26 rai = 1 hectare) resulted in an average yield increase of 73%. More work showed that applying bentonite to degraded sandy soils reduced the risk of crop failure during drought years.

In 2008, three years after the initial trials, IWMI scientists conducted a survey among 250 farmers in northeast Thailand, half of whom had applied bentonite to their fields. The average improvement for those using the clay addition was 18% higher than for non-clay users. Using the clay had enabled some farmers to switch to growing vegetables, which need more fertile soil. This helped to increase their income. The researchers estimated that 200 farmers in northeast Thailand and 400 in Cambodia had adopted the use of clays, and that a further 20,000 farmers were introduced to the new technique.

If the soil is too high in clay, adding gypsum, washed river sand and organic matter will balance the composition. Adding organic matter (like ramial chipped wood for instance) to soil which is depleted in nutrients and too high in sand will boost its quality.

REFERENCES

1. Voroney, R. Paul; Heck, Richard J. (2015). "The soil habitat". In Paul, Eldor A. Soil microbiology, ecology and biochemistry (4th ed.). Amsterdam, The Netherlands: Elsevier. doi:10.1016/B978-0-12-415955-6.00002-5. ISBN 978-0-12-415955-6.
2. Danoff-Burg, James A. "The terrestrial influence: geology and soils". Earth Institute Center for Environmental Sustainability. New York, New York: Columbia University Press. Retrieved 25 September 2016.
3. Taylor, Sterling A.; Ashcroft, Gaylen L. (1972). Physical edaphology: the physics of irrigated and nonirrigated soils. San Francisco, California: W.H. Freeman. ISBN 978-0716708186.
4. McCarthy, David F. (2006). Essentials of soil mechanics and foundations: basic geotechnics (7th ed.). Upper Saddle River, New Jersey: Prentice Hall. ISBN 978-0131145603.

5. Gilluly, James; Waters, Aaron Clement; Woodford, Alfred Oswald (1975). Principles of geology (4th ed.). San Francisco, California: W.H. Freeman. ISBN 978-0716702696.
6. Ponge, Jean-François (2015). "The soil as an ecosystem" (PDF). Biology and Fertility of Soils. 51 (6): 645–648. doi:10.1007/s00374-015-1016-1.
7. "Soil density". Argonne National Laboratory. Retrieved 25 September 2016.
8. Buol, Stanley W.; Southard, Randal J.; Graham, Robert C.; McDaniel, Paul A. (2011). Soil genesis and classification (6th ed.). Ames, Iowa: Wiley-Blackwell. ISBN 978-0470960608.
9. Retallack, Gregory J. (2008). Soils of the past: an introduction to paleopedology (2nd ed.). Hoboken, New Jersey: John Wiley & Sons. ISBN 978-0470698167.
10. "Glossary of Terms in Soil Science". Agriculture and Agri-Food Canada. Retrieved 9 October 2016.
11. Amundson, Ronald. "Soil preservation and the future of pedology" (PDF). Faculty of Natural Resources. Prince of Songkla University, Songkhla, Thailand. Retrieved 9 October 2016.
12. Simonson, Roy W. (1957). "What soils are". The yearbook of agriculture 1957(PDF) (1st ed.). Washington, D.C.: United States Government Printing Office.
13. Raloff, Janet (17 July 2008). "Dirt is not soil". ScienceNews. Retrieved 23 October 2016.
14. Pouyat, Richard; Groffman, Peter; Yesilonis, Ian; Hernandez, Luis (2002). "Soil carbon pools and fluxes in urban ecosystems" (PDF). Environmental Pollution. 116 (Supplement 1): S107–S118. doi:10.1016/S0269-7491(01)00263-9.
15. Davidson, Eric A.; Janssens, Ivan A. (2006). "Temperature sensitivity of soil carbon decomposition and feedbacks to climate change" (PDF). Nature. 440 (9 March 2006): 165–173. doi:10.1038/nature04514.
16. Powlson, David (2005). "Climatology: will soil amplify climate change?". Nature. 433 (20 January 2005): 204–205. doi:10.1038/433204a. (subscription required (help)).
17. Bradford, Mark A.; Wieder, William R.; Bonan, Gordon B.; Fierer, Noah; Raymond, Peter A.; Crowther, Thomas W. (2016). "Managing uncertainty in soil carbon feedbacks to climate change" (PDF). Nature Climate Change. 6 (27 July 2016): 751–758. doi:10.1038/nclimate3071.
18. Dominati, Estelle; Patterson, Murray; Mackay, Alec (2010). "A framework for classifying and quantifying the natural capital and ecosystem services of soils"(PDF). Ecological Economics. 69 (9): 1858–1868. doi:10.1016/j.ecolecon.2010.05.002.
19. Dykhuizen, Daniel E. (1998). "Santa Rosalia revisited: why are there so many species of bacteria?" (PDF). Antonie van Leeuwenhoek. 73 (1): 25–33. doi:10.1023/A:1000665216662.
20. Torsvik, Vigdis; Øvreås, Lise (2002). "Microbial diversity and function in soil: from genes to ecosystems" (PDF). Current Opinion in Microbiology. 5 (3):

240–245. doi:10.1016/S1369-5274(02)00324-7.

21. Raynaud, Xavier; Nunan, Naoise (2014). "Spatial ecology of bacteria at the microscale in soil" (PDF). PLOS ONE. 9 (1): e87217. doi:10.1371/journal.pone.0087217.
22. Whitman, William B.; Coleman, David C.; Wiebe, William J. (1998). "Prokaryotes: the unseen majority" (PDF). Proceedings of the National Academy of Sciences of the USA. 95 (12): 6578–6583. doi:10.1073/pnas.95.12.6578.
23. Schlesinger, William H.; Andrews, Jeffrey A. (2000). "Soil respiration and the global carbon cycle" (PDF). Biogeochemistry. 48 (1): 7–20. doi:10.1023/A:1006247623877.
24. Denmead, Owen Thomas; Shaw, Robert Harold (1962). "Availability of soil water to plants as affected by soil moisture content and meteorological conditions"(PDF). Agronomy Journal. 54 (5): 385–390. doi:10.2134/agronj1962.00021962005400050005x.
25. House, Christopher H.; Bergmann, Ben A.; Stomp, Anne-Marie; Frederick, Douglas J. (1999). "Combining constructed wetlands and aquatic and soil filters for reclamation and reuse of water" (PDF). Ecological Engineering. 12 (1–2): 27–38. doi:10.1016/S0925-8574(98)00052-4.
26. Van Bruggen, Ariena H.C.; Semenov, Alexander M. (2000). "In search of biological indicators for soil health and disease suppression" (PDF). Applied Soil Ecology. 15 (1): 13–24. doi:10.1016/S0929-1393(00)00068-8.
27. "A citizen's guide to monitored natural attenuation" (PDF). Retrieved 27 November 2016.
28. Linn, Daniel Myron; Doran, John W. (1984). "Effect of water-filled pore space on carbon dioxide and nitrous oxide production in tilled and nontilled soils" (PDF). Soil Science Society of America Journal. 48 (6): 1267–1272. doi:10.2136/sssaj1984.03615995004800060013x.
29. Miller, Raymond W.; Donahue, Roy Luther (1990). Soils: an introduction to soils and plant growth. Upper Saddle River, New Jersey: Prentice Hall. ISBN 978-0138202262.
30. Bot, Alexandra; Benites, José (2005). The importance of soil organic matter: key to drought-resistant soil and sustained food and production (PDF). Rome, Italy: Food and Agriculture Organization of the United Nations. ISBN 92-5-105366-9. Retrieved 27 November 2016.
31. Barber, Stanley A. (1995). "Chemistry of soil-nutrient associations". In Barber, Stanley A. Soil nutrient bioavailability: a mechanistic approach (2nd ed.). New York, New York: John Wiley & Sons. pp. 9–48. ISBN 978-0-471-58747-7.
32. Möller, Detlev (2015). "Chemistry under environmental conditions". In Möller, Detlev. Chemistry for environmental scientists (1st ed.). Berlin, Germany: Walter de Gruyter. pp. 7–52. doi:10.1515/9783110410013-004. ISBN 978-3-11-041001-3.
33. "Arizona Master Gardener Manual". Cooperative Extension, College of Agriculture, University of Arizona. Retrieved 4 December 2016.

34. Vannier, Guy (1987). "The porosphere as an ecological medium emphasized in Professor Ghilarov's work on soil animal adaptations". Biology and Fertility of Soils. 3 (1): 39–44. doi:10.1007/BF00260577.
35. Torbert, H. Allen; Wood, Wes (1992). "Effect of soil compaction and water-filled pore space on soil microbial activity and N losses" (PDF). Communications in Soil Science and Plant Analysis. 23 (11): 1321–1331. doi:10.1080/00103629209368668.
36. Simonson 1957, p. 17.
37. Fookes, Peter G. (1997). Tropical residual soils. London, United Kingdom: Geological Society of London. ISBN 978-1897799383.
38. "Soil and water". Food and Agriculture Organization of the United Nations. Retrieved 11 December 2016.
39. "Sources. Negative Charge:". Jan.ucc.nau.edu. Retrieved 7 November 2012.
40. Donahue, Miller & Shickluna 1977, p. 4.
41. Kellogg 1957, p. 1.
42. Brady, Nyle (1984). The Nature and Properties of Soils (9th ed.). USA: Macmillan Publishing Co. pp. 4–7. ISBN 0023133406.
43. Kellogg 1957, p. 3.
44. Kellogg 1957, p. 2.
45. Kellogg 1957, p. 4.
46. "nitrogen fixation." Encyclopædia Britannica. Encyclopædia Britannica 2009 Deluxe Edition. Chicago: Encyclopædia Britannica, 2009.
47. Kellogg 1957, pp. 1–4.
48. Michael E. Ritter. Factors Affecting Soil Development, Soil Systems, The Physical Environment: an Introduction to Physical Geography, University of Wisconsin, Stevens Point, 1 October 2009, retrieved 3 January 2012.
49. "Soils". Natural Resource Conservation Service. United States Department of Agriculture. Retrieved 26 May 2013.
50. Van Schöll, Laura; Smits, Mark M.; Hoffland, Ellis (2006). "Ectomycorrhizal weathering of the soil minerals muscovite and hornblende". New Phytologist. 171 (4): 805–814. doi:10.1111/j.1469-8137.2006.01790.x. PMID 16918551.
51. Donahue, Miller & Shickluna 1977, pp. 20–21.
52. Donahue, Miller & Shickluna 1977, p. 21.
53. Donahue, Miller & Shickluna 1977, p. 24.
54. Gilluly, Waters, Woodford (1975). Principles of Geology (4th ed.). USA: W.H. Freeman. p. 215. ISBN 978-0716702696.
55. Donahue, Miller & Shickluna 1977, pp. 28–31.
56. Donahue, Miller & Shickluna 1977, pp. 31–33.
57. H.B. Milford, A.J.E. McGaw and K.J. Nixon, Soil Data Entry Handbook for the NSW Soil and Land Information System (SALIS), 3rd ed., New South Wales Department of Land and Water Conservation Resource Information Systems

Group, Parramatta, 2001, pdf pp. 30–32.

58. Gove Hambidge, "Climate and Man—A Summary", in Erwin Raisz, U.S. Department of Agriculture, Climate And Man, Part One, Yearbook of Agriculture 1941, Repr. Honolulu: University Press of the Pacific, 2004, ISBN 978-1-4102-1538-3, pp. 1–66, p. 27.
59. Donahue, Miller & Shickluna 1977, p. 35.
60. Copley, Jon (25 August 2005). "Millions of bacterial species revealed underfoot". New Scientist. Retrieved 19 April 2010.
61. Amber Dance (2008). "Soil ecology: What lies beneath". Nature. 455 (7214): 724–5. doi:10.1038/455724a. PMID 18843336.
62. Roesch LF, Fulthorpe RR, Riva A, Casella G, Hadwin AK, Kent AD, Daroub SH, Camargo FA, Farmerie WG, Triplett EW; Fulthorpe; Riva; Casella; Hadwin; Kent; Daroub; Camargo; Farmerie; Triplett (2007). "Pyrosequencing enumerates and contrasts soil microbial diversity". The ISME Journal. 1 (4): 283–90. doi:10.1038/ismej.2007.53. PMC 2970868 . PMID 18043639.
63. Gans J, Wolinsky M, Dunbar J.; Wolinsky; Dunbar (2005). "Computational improvements reveal great bacterial diversity and high metal toxicity in soil". Science. 309 (5739): 1387–90. Bibcode:2005Sci...309.1387G. doi:10.1126/science.1112665. PMID 16123304.
64. Simonson 1957, pp. 20, 21.
65. Donahue, Miller & Shickluna 1977, p. 26.
66. Nyle C. Brady & Ray R. Weil (2009). Elements of the Nature and Properties of Soils (3rd Edition). Prentice Hall. ISBN 9780135014332.
67. R. B. Brown (September 2003). "Soil Texture" (PDF). Fact Sheet SL-29. University of Florida, Institute of Food and Agricultural Sciences. Retrieved 8 July 2008.
68. Russell 1957, pp. 32–33.
69. Flemming 1957, p. 331.
70. Donahue, Miller & Shickluna 1977, p. 53.
71. Soil Survey Division Staff (1993). "Soil Structure". Handbook 18. Soil survey manual. Soil Conservation Service. U.S. Department of Agriculture. Archived from the original on 16 March 2008. Retrieved 8 July 2008.
72. Donahue, Miller & Shickluna 1977, pp. 55–56.
73. "Soil Structure - Physical Properties". The Cooperative Soil Survey. Retrieved 19 October 2012.
74. Donahue, Miller & Shickluna 1977, pp. 59–61.
75. Donahue, Miller & Shickluna 1977, p. 60.
76. "Physical aspects of crop productivity". www.fao.org. FAO. Retrieved 2 July 2016.
77. Donahue, Miller & Shickluna 1977, pp. 62–63.

78. "Soils - Part 2: Physical Properties of Soil and Soil Water". Retrieved 19 October 2012.

79. Johnson, T.A., T.R. Ellsworth, R.J.M. Hudson, and G.K. Sims. 2013. Diffusion Limitation for Atrazine Biodegradation in Soil. Advances in Microbiology. 3(5): 412-420.

80. Donahue, Miller & Shickluna 1977, pp. 62–63,565–567.

81. Taylor, H.M. 1983. Managing root systems for efficient water use: an overview, p. 87–113 in Taylor, H.H. et al. (Eds.). Limitations to Efficient Water Use in Crop Production. ASA, CCSA, and SSSA, Madison WI.

82. Gill, D. 1975. Influence of white spruce trees on permafrost-table microtopography, Mackenzie River Delta. Can. J. Earth Sci. 12(2):263–272.

83. Mityga, H.G.; Lanphear, F.O. 1971. Factors influencing the cold hardiness of Taxus cuspidata roots. J. Amer. Soc. Hort. Sci. 96:83–86.

84. Havis, J.R. 1976. Root hardiness of woody ornamentals. HortScience 11:385–386.

85. Lyr, H.; Hoffmann, G. 1967. Growth rates and growth periodicity of tree roots. p. 181–236 in Romberger, J.A.; Mikola, P. (Eds.). International Review of Forest Research, Vol. 2, Academic Press, New York NY.

86. Chalupa, V.; Fraser, D.A. 1968. Effect of soil and air temperature on soluble sugars and growth of white spruce (Picea glauca) seedlings. Can. J. Bot. 46:65–69.

87. Heninger, R.L.; White, D.P. 1974. Tree seedling growth at different soil temperatures. For. Sci. 20:363–367.

88. Ritchie, G.A.; Dunlap, J.R. 1980. Root growth potential: its development and expression in forest tree seedlings. New Zealand J. For. Sci. 10:218–248.

89. Binder, W.D.; Spittlehouse, D.L.; Draper, D.A. 1988. Post-planting studies of the physiology of white spruce 1984–1985, E.P. 966. B.C. Min. For., Res. Branch, Victoria BC, Progr. Rep. 5, unpubl. 85 p. [Coates et al. 1994 give date as 1987, Binder bibliog at R986 gives 1988]

90. Landhäusser, S.M.; DesRochers, A.; and Lieffers, V.J. 2001. A comparison of growth and physiology in Picea glauca and Populus tremuloides at different soil temperatures. Can. J. For. Res. 31:1922–1929.

91. Tryon, P.R.; Chapin, F.S. 1983. Temperature control over root growth and root biomass in taiga forest trees. Can. J. For. Res. 13(5):827 833.

92. Landhäusser, S.M.; Silins, U.; Lieffers, V.J.; Liu, W. 2003. Response of Populus tremuloides, Populus balsamifera, Betula papyrifera and Picea glauca seedlings to low soil temperature and water-logged soil conditions. Scan. J. For. Res. 18:391–400. (Cited Green 2004)

93. Turner, N.C.; Jarvis, P.G. 1975. Photosynthesis in Sitka spruce (Picea sitchensis Bong.). IV. Response to soil temperature. J. Appl. Ecol. 12:561–576.

94. Day, T.A.; DeLucia, E.H.; Smith, W.K. 1990. Effect of soil temperature on stem flow, shoot gas exchange and water potential of Picea engelmannii (Parry) during snowmelt. Oecologia 84(4):474–481.

95. Green, D.S. 2004. Describing condition-specific determinants of competition in boreal and sub-boreal mixedwood stands. For. Chron. 80(6):736–742.
96. Donahue, Miller & Shickluna 1977, p. 71.
97. "Arizona Master Gardener Manual". Cooperative Extension, College of Agriculture, University of Arizona. p. Chapter 2, pp 4–8. Retrieved 27 May 2013.
98. "The Color of Soil". United States Department of Agriculture - Natural Resources Conservation Service. Archived from the original on 16 March 2008. Retrieved 8 July 2008.
99. "Electrical Design, Cathodic Protection". United States Army Corps of Engineers. 22 April 1985. Archived from the original on 12 June 2008. Retrieved 2 July 2008.
100. "The why and how to testing the Electrical Conductivity of Soils | Resources". Retrieved 19 December 2010.
101. R. J. Edwards (15 February 1998). "Typical Soil Characteristics of Various Terrains". Retrieved 2 July 2008.
102. Donahue, Miller & Shickluna 1977, p. 72.
103. Wadleigh 1957, p. 48.
104. Richards & Richards 1957, p. 50.
105. Richards & Richards 1957, p. 56.
106. Wadleigh 1957, p. 39.
107. Richards & Richards 1957, p. 52.
108. Donahue, Miller & Shickluna 1977, pp. 72-74.
109. "CHAPTER 2 - SOIL AND WATER". Fao.org. Retrieved 7 November 2012.
110. Donahue, Miller & Shickluna 1977, pp. 72–75.
111. Wadleigh 1957, p. 47.
112. Donahue, Miller & Shickluna 1977, pp. 75–76.
113. Donahue, Miller & Shickluna 1977, pp. 76–77.
114. Donahue, Miller & Shickluna 1977, p. 80.
115. Donahue, Miller & Shickluna 1977, p. 85.
116. Donahue, Miller & Shickluna 1977, p. 86.
117. Donahue, Miller & Shickluna 1977, p. 88.
118. Brehm, Denise (11 December 2008). "CEE researchers explain mystery of gravity fingers". MIT Department of Civil & Environmental Engineering. MIT. Retrieved 31 October 2012.
119. "Urban Trees Enhance Water Infiltration". Fisher, Madeline. The American Society of Agronomy. 17 November 2008. Retrieved 31 October 2012.
120. "Major floods recharge aquifers". University of New South Wales Science. 24 January 2011. Retrieved 31 October 2012.
121. Donahue, Miller & Shickluna 1977, p. 90.

122. Donahue, Miller & Shickluna 1977, p. 91.
123. Donahue, Miller & Shickluna 1977, p. 92.
124. Wadleigh 1957, p. 46.
125. Donahue, Miller & Shickluna 1977, p. 94.
126. Donahue, Miller & Shickluna 1977, pp. 97–99.
127. Russell 1957, pp. 35–36.
128. http://atmo.tamu.edu/class/geos489/lecture6/soiltracegasrev.pdf
129. College of Tropical Agriculture and Human Resources. "Soil Mineralogy". www.ctahr.hawaii.edu. University of Hawai'i. Retrieved 7 August 2014.
130. Russell, E. Walter (1973). Soil conditions and plant growth. (10th ed.). London: Longman. pp. 67–70. ISBN 0582440483.
131. Donahue, Miller & Shickluna 1977, pp. 101–102.
132. Simonson 1957, p. 19.
133. Donahue, Miller & Shickluna 1977, p. 102.
134. Russell 1957, p. 33.
135. Donahue, Miller & Shickluna 1977, pp. 102–107.
136. Donahue, Miller & Shickluna 1977, pp. 101–107.
137. Donahue, Miller & Shickluna 1977, p. 107.
138. Donahue, Miller & Shickluna 1977, p. 108.
139. Russell 1957, pp. 33, 34.
140. Coleman & Mehlich 1957, p. 74.
141. Donahue, Miller & Shickluna 1977, pp. 108–110.
142. Dean 1957, p. 82.
143. Allison 1957, p. 90.
144. Reitemeier 1957, p. 103.
145. Donahue, Miller & Shickluna 1977, p. 110.
146. Coleman & Mehlich 1957, p. 73.
147. Holmes & Brown 1957, p. 112.
148. Donahue, Miller & Shickluna 1977, p. 111.
149. Olsen & Fried 1957, p. 96.
150. Reitemeier 1957, p. 101.
151. Donahue, Miller & Shickluna 1977, pp. 103–112.
152. Simonson 1957, pp. 18, 21–24, 29.
153. Russell 1957, pp. 32, 35.
154. Donahue, Miller & Shickluna 1977, p. 112.
155. Russell 1957, p. 35.
156. Allaway 1957, p. 69.

157. Lehmann, J. "Terra Preta de Indio". University of Cornell, Dept. of Crop and Soil Sciences. Retrieved 30 March 2013.
158. Donahue, Miller & Shickluna 1977, p. 103–106.
159. Donahue, Miller & Shickluna 1977, p. 114.
160. Donahue, Miller & Shickluna 1977, pp. 115–116.
161. Chang, Raymond (1984). Chemistry. Random House, Inc. p. 424. ISBN 0-394-32983-X.
162. Donahue, Miller & Shickluna 1977, pp. 116–117.
163. Donahue, Miller & Shickluna 1977, pp. 116–119.
164. Donahue, Miller & Shickluna 1977, pp. 119–120.
165. Donahue, Miller & Shickluna 1977, pp. 120–121.
166. Dean 1957, p. 80.
167. Russel 1957, pp. 123–125.
168. Dean 1957, pp. 80, 81.
169. Roy, R.N.; Finck, A.; Blair, G.J.; Tandon, H.L.S. (2006). "Chapter 4: Soil fertility and crop production". Plant nutrition for food security: a guide for integrated nutrient management (PDF). Rome: Food and Agriculture Organization of the United Nations. pp. 43–90. ISBN 92-5-105490-8. Retrieved 20 June 2016.
170. Donahue, Miller & Shickluna 1977, pp. 123–131.
171. Donahue, Miller & Shickluna 1977, p. 125.
172. Donahue, Miller & Shickluna 1977, p. 126.
173. "Lecture 22". Northern Arizona University. Retrieved 22 March 2013.
174. Donahue, Miller & Shickluna 1977, pp. 123–128.
175. Wadleigh 1957, p. 41.
176. Broadbent 1957, p. 153.
177. Donahue, Miller & Shickluna 1977, p. 128.
178. Allison 1957, pp. 85–94.
179. Broadbent 1957, pp. 152–155.
180. Donahue, Miller & Shickluna 1977, pp. 128–131.
181. Donahue, Miller & Shickluna 1977, pp. 129–130.
182. Donahue, Miller & Shickluna 1977, p. 145.
183. Donahue, Miller & Shickluna 1977, pp. 128–129.
184. Allison 1957, p. 87.
185. Donahue, Miller & Shickluna 1977, p. 130.
186. Donahue, Miller & Shickluna 1977, p. 131.
187. Donahue, Miller & Shickluna 1977, pp. 134–135.
188. Reitemeier 1957, pp. 101–104.

189. Donahue, Miller & Shickluna 1977, pp. 135–136.
190. Donahue, Miller & Shickluna 1977, p. 136.
191. Jordan & Reisenauer 1957, pp. 107.
192. Holmes & Brown 1957, pp. 111.
193. Sherman 1957, pp. 135.
194. Seatz & Jurinak 1957, pp. 115.
195. Reuther 1957, pp. 128.
196. Russel 1957, pp. 121.
197. Stout & Johnson 1957, pp. 146.
198. Stout & Johnson 1957, pp. 141.
199. Donahue, Miller & Shickluna 1977, pp. 136–137.
200. Stout & Johnson 1957, pp. 107.
201. Pimentel, D.; et al. (1995). "Environmental and economic costs of soil erosion and conservation benefits". Science. 267 (24): 1117–22. Bibcode:1995Sci...267.1117P. doi:10.1126/science.267.5201.1117. PMID 17789193.
202. Foth, Henry D. (1984). Fundamentals of soil science. New York: Wiley. p. 151. ISBN 0-471-88926-1.
203. Gilluly, Waters, Woodford (1975). Principles of Geology (4th ed.). USA: W.H. Freeman. p. 216. ISBN 978-0716702696.
204. Verkaik, Eric; Jongkind, Anne G.; Berendse, Frank (2006). "Short-term and long-term effects of tannins on nitrogen mineralization and litter decomposition in kauri (Agathis australis (D. Don) Lindl.) forests". Plant and Soil. 287: 337. doi:10.1007/s11104-006-9081-8.
205. Fierer, N.; Schimel, Joshua P.; Cates, Rex G.; Zou, Jiping (2001). "Influence of balsam poplar tannin fractions on carbon and nitrogen dynamics in Alaskan taiga floodplain soils". Soil Biology and Biochemistry. 33 (12–13): 1827. doi:10.1016/S0038-0717(01)00111-0.
206. Wagai, Rota; Mayer, Lawrence M.; Kitayama, Kanehiro; Knicker, Heike (2008). "Climate and parent material controls on organic matter storage in surface soils: A three-pool, density-separation approach". Geoderma. 147: 23–33. doi:10.1016/j.geoderma.2008.07.010.
207. Minayeva, T. Yu.; Trofimov, S. Ya.; Chichagova, O.A.; Dorofeyeva, E.I.; Sirin, A.A.; Glushkov, I.V.; Mikhailov, N.D.; Kromer, B. (2008). "Carbon accumulation in soils of forest and bog ecosystems of southern Valdai in the Holocene". Biology Bulletin. 35 (5): 524–532. doi:10.1134/S1062359008050142.
208. Sanchez, Pedro A. (1976). Properties and management of soils in the tropics. New York: Wiley. ISBN 0471752002.
209. Paul, E.A. (1997). Soil organic matter in temperate agroecosystems : long-term experiments in North America. Boca Raton: CRC Press. p. 80. ISBN 9780849328022.

210. Retallack, G. J. (1990). Soils of the past : an introduction to paleopedology. Boston: Unwin Hyman. p. 32. ISBN 978-0-04-445757-2.

211. Buol, S.W. (1990). Soil genesis and classification. Ames, Iowa: Iowa State University Press. p. 36. doi:10.1081/E-ESS. ISBN 0-8138-2873-2.

212. IUSS Working Group WRB (2014). World Reference Base for Soil Resources 2014. International soil classification system for naming soils and creating legends for soil maps (PDF) (3rd ed.). Rome: FAO. ISBN 978-92-5-108370-3. Retrieved 29 August 2014.

213. bhttp://eusoils.jrc.ec.europa.eu/esdb_archive/eusoils_docs/other/EUR23439.pdfSoils of the European Union by the EU Institute for Environment and Sustainability. Accessed on 8 October 2013

214. The Soil Orders, Department of Environmental Sciences, University of Virginia, retrieved 23 October 2012.

215. Donahue, Miller & Shickluna 1977, pp. 411–432.

216. Soil Survey Staff (1999). Soil taxonomy: A Basic System of Soil Classification for Making and Interpreting Soil Surveys. 2nd edition. Natural Resources Conservation Service. U.S. Department of Agriculture Handbook 436. (PDF). USDA: United States Dept. of Agriculture, Naturel Resources Conservation Service. Retrieved 10 March 2013.

217. The Twelve Soil Orders: Soil Taxonomy, Soil & Land Resources Division, College of Agricultural and Life Sciences, University of Idaho

218. Donahue, Miller & Shickluna 1977, p. 409.

219. Donahue, Miller & Shickluna 1977, pp. 409-410.

220. Donahue, Miller & Shickluna 1977, p. 410.

221. Leake, Simon; Haege, Elke (2014). Soils for Landscape Development. CSIRO Publishing. ISBN 9780643109643.

222. Ponge, Jean-François (2003). "Humus forms in terrestrial ecosystems: a framework to biodiversity" (PDF). Soil Biology and Biochemistry. 35 (7): 935–945. doi:10.1016/S0038-0717(03)00149-4.

223. De Deyn, Gerlinde B.; Van der Putten, Wim H. (2005). "Linking aboveground and belowground diversity". Trends in Ecology & Evolution. 20 (11): 625–633. doi:10.1016/j.tree.2005.08.009. PMID 16701446.

224. Hansen, J.; et al. (2008). "Target atmospheric CO2: Where should humanity aim?". Open Atmospheric Science Journal. 2: 217–31. arXiv:0804.1126 . Bibcode:2008OASJ....2..217H. doi:10.2174/1874282300802010217

225. Lal, R. (11 June 2004). "Soil Carbon Sequestration Impacts on Global Climate Change and Food Security". Science. 304 (5677): 1623–1627. Bibcode:2004Sci...304.1623L. doi:10.1126/science.1097396. PMID 15192216.

226. Blakeslee, Thomas R. (24 February 2010). "Greening Deserts for Carbon Credits". Renewable Energy World.com. Retrieved 23 October 2012.

227. Setz, EZF; Enzweiler J; Solferini VN; Amendola MP; Berton RS (1999). "Geophagy

in the golden-faced saki monkey (Pithecia pithecia chrysocephala) in the Central Amazon". Journal of Zoology. 247 (1): 91–103. doi:10.1111/j.1469-7998.1999.tb00196.x. (subscription required (help)).

228. Kohne, John Maximilian; Koehne, Sigrid; Simunek, Jirka (2009). "A review of model applications for structured soils: a) Water flow and tracer transport". Journal of Contaminant Hydrology. 104 (1–4): 4–35. Bibcode:2009JCHyd.104....4K. doi:10.1016/j.jconhyd.2008.10.002. PMID 19012994.
229. Diplock, EE; Mardlin DP; Killham KS; Paton GI (2009). "Predicting bioremediation of hydrocarbons: laboratory to field scale". Environmental Pollution. 157 (6): 1831–1840. doi:10.1016/j.envpol.2009.01.022. PMID 19232804.
230. Moeckel, Claudia; Nizzetto, Luca; Di Guardo, Antonio; Steinnes, Eiliv; Freppaz, Michele; Filippa, Gianluca; Camporini, Paolo; Benner, Jessica; Jones, Kevin C. (2008). "Persistent organic pollutants in boreal and montane soil profiles: distribution, evidence of processes and implications for global cycling". Environmental Science and Technology. 42 (22): 8374–8380. Bibcode:2008EnST...42.8374M. doi:10.1021/es801703k. PMID 19068820.
231. Rezaei, Khalil; Guest, Bernard; Friedrich, Anke; Fayazi, Farajollah; Nakhaei, Mohamad; Aghda, Seyed Mahmoud Fatemi; Beitollahi, Ali (2009). "Soil and sediment quality and composition as factors in the distribution of damage at the December 26, 2003, Bam area earthquake in SE Iran (M (s)=6.6)". Journal of Soils and Sediments. 9: 23–32. doi:10.1007/s11368-008-0046-9.
232. Johnson, D.L.; Ambrose, S.H.; Bassett, T.J.; Bowen, M.L.; Crummey, D.E.; Isaacson, J.S.; Johnson, D.N.; Lamb, P.; Saul, M.; Winter-Nelson, A. E. (1997). "Meanings of environmental terms". Journal of Environmental Quality. 26 (3): 581–589. doi:10.2134/jeq1997.00472425002600030002x.
233. Jones, j. a. a. (1976). "Soil piping and stream channel initiation". Water Resources Research. 7 (3): 602–610. Bibcode:1971WRR.....7..602J. doi:10.1029/WR007i003p00602.
234. Dooley, Alan (June 2006). "Sandboils 101: Corps has experience dealing with common flood danger". Engineer Update. US Army Corps of Engineers. Archived from the original on 18 April 2008. Retrieved 14 May 2008.
235. ILRI (1989). "Effectiveness and Social/Environmental Impacts of Irrigation Projects: a Review" (pdf). In: Annual Report 1988 of the International Institute for Land Reclamation and Improvement (ILRI). Wageningen, The Netherlands. pp. 18–34.
236. Drainage Manual: A Guide to Integrating Plant, Soil, and Water Relationships for Drainage of Irrigated Lands. Interior Dept., Bureau of Reclamation. 1993. ISBN 0-16-061623-9.
237. "Free articles and software on drainage of waterlogged land and soil salinity control". Retrieved 28 July 2010.
238. Improving soils and boosting yields in Thailand" (PDF). Success Stories.

International Water Management Institute (2). 2010. doi:10.5337/2011.0031.

239. Provide for your garden's basic needs ... and the plants will take it from there". USA Weekend. 10 March 2011.

CHAPTER 4

Soil: Chemical and Physical Characters

Chapter Contents

Soil chemistry is the study of the chemical characteristics of soil. Soil chemistry is affected by mineral composition, organic matter and environmental factors. Soil - (i) The unconsolidated mineral or organic material on the immediate surface of the Earth that serves as a natural medium for the growth of land plants. (ii) The unconsolidated mineral or organic matter on the surface of the Earth that has been subjected to and shows effects of genetic and environmental factors of: climate (including water and temperature effects), and macro- and microorganisms, conditioned by relief, acting on parent material over a period of time. A product-soil differs from the material from which it is derived in many physical, chemical,biological, and morphological properties and characteristics. This definition is from Soil

INTRODUCTION

Soil makes up the outermost layer of our planet and is formed from rocks and decaying plants and Animals. Soil has varying amounts of organic matter (living and dead organisms), minerals, and nutrients. • An average soil sample is 45 percent minerals, 25 percent water, 25 percent air, and five percent organic matter. Different-sized mineral particles, such as sand, silt, and clay, give soil its texture. • Topsoil is the most productive soil layer. • Ten tonnes of topsoil spread evenly over a hectare is only as thick as a one Euro coin. • Natural processes can take more than 500 years to form 2 centimeters of topsoil. • In some cases, 5 tonnes of animal life can live in one hectare of soil. • Fungi and bacteria help break down organic matter in the soil. • Earthworms digest organic matter, recycle nutrients, and make the surface soil richer. • Roots loosen the soil, allowing oxygen to penetrate. This benefits animals living in the soil. They also benefit roots which require oxygen themselves. They further hold soil together and help prevent erosion. • A fully functioning soil reduces the risk of floods and protects underground water supplies by neutralizing or filtering out potential pollutants and storing as much as 3750 tonnes of water per hectare. • Scientists have found 10,000 types of soil in Europe and about 70,000 types of soil in the United States. • Soil stores 10% of the world's carbon dioxide emissions.

ENVIRONMENTAL SOIL CHEMISTRY

A knowledge of environmental soil chemistry is paramount to predicting the fate of contaminants, as well as the processes by which they are initially released into the soil. Once a chemical is exposed to the soil environment myriad chemical reactions can occur that may increase or decrease contaminant toxicity. These reactions include adsorption/desorption, precipitation, polymerization, dissolution, complexation and oxidation/reduction. These reactions are often disregarded by scientists and engineers involved with environmental remediation. Understanding these processes enable us to better predict the fate and toxicity of contaminants and provide the knowledge to develop scientifically correct, and cost-effective remediation strategies.

SOIL PH

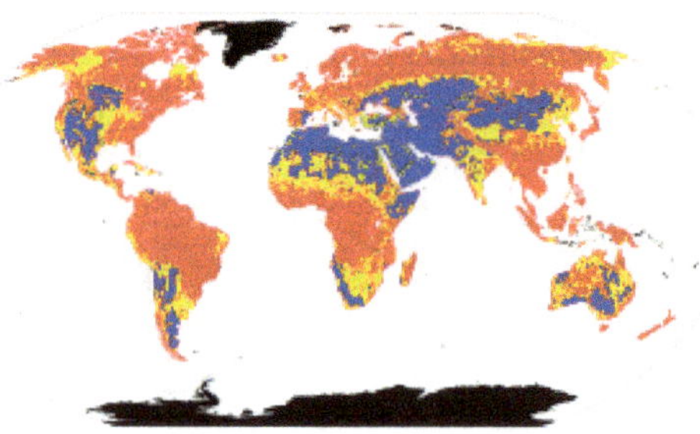

Global variation in soil pH. **Red** = acidic soil. **Yellow** = neutral soil. **Blue** = alkaline soil. **Black** = no data.

The **soil pH** is a measure of the acidity or alkalinity in soils. pH is defined as the negative logarithm (base 10) of the activity of hydronium ions (H^+ or, more precisely, $H3O^+aq$) in a solution. In water, it normally ranges from -1 to 14, with 7 being neutral. A pH below 7 is acidic and above 7 is alkaline. Soil pH is considered a master variable in soils as it controls many chemical processes that take place. It specifically affects plant nutrient availability by controlling the chemical forms of the nutrient. The optimum pH range for most plants is between 5.5 and 7.0, however many plants have adapted to thrive at pH values outside this range.

Classification of soil pH Ranges

The United States Department of Agriculture Natural Resources Conservation Service, formerly Soil Conservation Service classifies soil pH ranges as follows:

Denomination	pH range
Ultra acidic	< 3.5
Extremely acidic	3.5–4.4
Very strongly acidic	4.5–5.0
Strongly acidic	5.1–5.5
Moderately acidic	5.6–6.0
Slightly acidic	6.1–6.5
Neutral	6.6–7.3
Slightly alkaline	7.4–7.8
Moderately alkaline	7.9–8.4
Strongly alkaline	8.5–9.0
Very strongly alkaline	> 9.0

Sources of soil pH

Sources of acidity

Acidity in soils comes from H^+ and Al^{3+} ions in the soil solution and sorbed to soil surfaces. While pH is the measure of H^+ in solution, Al^{3+} is important in acid soils because between pH 4 and 6, Al^{3+} reacts with water (H_2O) forming $AlOH^{2+}$, and $Al(OH)_2^+$, releasing extra H^+ ions. Every Al^{3+}ion can create 3 H^+ ions. Many other processes contribute to the formation of acid soils including rainfall, fertilizer use, plant root activity and the weathering of primary and secondary soil minerals. Acid soils can also be caused by pollutants such as acid rain and mine spoilings.

- Rainfall: Acid soils are most often found in areas of high rainfall. Excess rainfall leaches base cation from the soil, increasing the percentage of Al^{3+} and H^+ relative to other cations. Additionally, rainwater has a slightly acidic pH of 5.7 due to a reaction with CO_2 in the atmosphere that forms carbonic acid.

- Fertilizer use: Ammonium (NH_4^+) fertilizers react in the soil in a process called nitrification to form nitrate (NO_3^-), and in the process release H^+ ions.
- Plant root activity: Plants take up nutrients in the form of ions (NO_3^-, NH_4^+, Ca^{2+}, $H_2PO_4^-$, etc.), and often, they take up more cations than anions. However plants must maintain a neutral charge in their roots. In order to compensate for the extra positive charge, they will release H^+ ions from the root. Some plants will also exude organic acids into the soil to acidify the zone around their roots to help solubilize metal nutrients that are insoluble at neutral pH, such as iron (Fe).
- Weathering of minerals: Both primary and secondary minerals that compose soil contain Al. As these minerals weather, some components such as Mg, Ca, and K, are taken up by plants, others such as Si are leached from the soil, but due to chemical properties, Fe and Al remain in the soil profile. Highly weathered soils are often characterized by having high concentrations of Fe and Al oxides.
- Acid rain: When atmospheric water reacts with sulfur and nitrogen compounds that result from industrial processes, the result can be the formation of sulfuric and nitric acid in rainwater. However the amount of acidity that is deposited in rainwater is much less, on average, than that created through agricultural activities.
- Mine spoil: Severely acidic conditions can form in soils near mine spoils due to the oxidation of pyrite.
- Potential acid sulfate soils naturally formed in waterlogged coastal and estuarine environments can become highly acidic when drained or excavated.
- Decomposition of organic matter by microorganisms releases CO_2 which when mixed with soil water can form carbonic acid (H_2CO_3).

Sources of Alkalinity

Alkaline soils have a high saturation of base cations (K^+, Ca^{2+}, Mg^{2+} and Na^+). This is due to an accumulation of soluble salts which are classified as either saline soil, sodic soil, saline-sodic soil or alkaline soil. All saline and sodic soils have high salt concentrations, with saline soils being dominated by calcium and magnesium salts and sodic soils being dominated by sodium. Alkaline soils are characterized by the presence of carbonates. Soil in areas with limestone near the surface are alkaline from the calcium carbonate in limestone constantly mixing with the soil.Groundwater sources in these areas contain dissolved limestone.

Effect of soil pH on Plant Growth

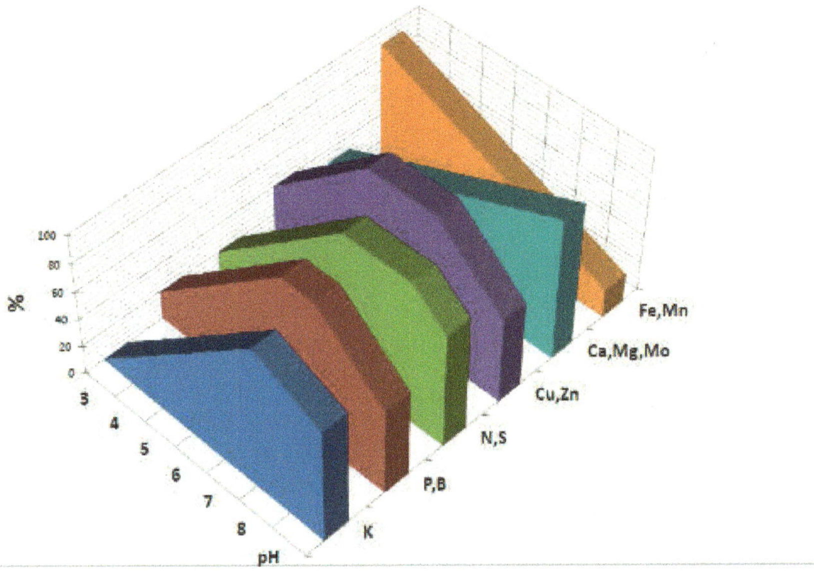

Nutrient availability in relation to soil pH.

Acid affected soils

Plants grown in acid soils can experience a variety of symptoms including aluminium (Al), hydrogen (H), and/or manganese (Mn) toxicity, as well as nutrient deficiencies of calcium (Ca) and magnesium (Mg).

Aluminium toxicity is the most widespread problem in acid soils. Aluminium is present in all soils, but dissolved Al^{3+} is toxic to plants; Al^{3+} is most soluble at low pH, above pH 5.2 little Al is in soluble form in most soils. Aluminium is not a plant nutrient, and as such, is not actively taken up by the plants, but enters plant roots passively through osmosis. Aluminium inhibits root growth; lateral roots and root tips become thickened and roots lack fine branching; root tips may turn brown. In the root, Al has been shown to interfere with many physiological processes including the uptake and transport of calcium and other essential nutrients, cell division, cell wall formation, and enzyme activity.

Below pH 4, H^+ ions themselves damage root cell membranes.

In soils with high content of manganese-containing minerals, Mn toxicity can become a problem at pH 5.6 and lower. Manganese, like aluminium, becomes increasingly soluble as pH drops, and Mn toxicity symptoms can be seen at pH levels below 5.6. Manganese is an essential plant nutrient, so plants transport Mn into leaves. Classic symptoms of Mn toxicity are crinkling or cupping of leaves.

Nutrient Availability in Relation to Soil pH

Nutrients needed in large amounts by plants are referred to as macronutrients and include nitrogen (N), phosphorus (P), potassium (K), calcium (Ca), magnesium (Mg) and sulfur (S). Elements that plants need in trace amounts are called trace nutrients or micronutrients. Trace nutrients are not major components of plant tissue but are essential for growth. They include iron (Fe), manganese (Mn), zinc (Zn), copper (Cu), cobalt (Co), molybdenum (Mo), and boron (B). Both macronutrient and micronutrient availability are affected by soil pH. In slightly to moderately alkaline soils, molybdenum and macronutrient (except for phosphorus) availability is increased, but P, Fe, Mn, Zn Cu, and Co levels are reduced and may adversely affect plant growth. In acidic soils, micronutrient availability (except for Mo and Bo) is increased. Nitrogen is supplied as ammonium (NH_4) or nitrate (NO_3) by nitrogen fixation or fertilizer amendments, and dissolved N will have the highest concentrations in soil with pH 6–8. Concentrations of available N are less sensitive to pH than concentration of available P. In order for P to be available for plants, soil pH needs to be in the range 6.0 and 7.5. If pH is lower than 6, P starts forming insoluble compounds with iron (Fe) and aluminium (Al) and if pH is higher than 7.5 P starts forming insoluble compounds with calcium (Ca). Most nutrient deficiencies can be avoided between a pH range of 5.5 to 6.5, provided that soil minerals and organic matter contain the essential nutrients to begin with.

Water availability in relation to soil pH

Determining pH

Methods of determining pH include:

- Observation of soil profile: Certain profile characteristics can be indicators of either acid, saline, or sodic conditions. Strongly acidic soils often have poor incorporation of the organic surface layer with the underlying mineral layer. The mineral horizons are distinctively layered in many cases, with a pale eluvial (E) horizon beneath the organic surface; this E is underlain by a darker B horizon in a classic podzol horizon sequence. This is a very rough gauge of acidity as there is no correlation between thickness of the E and soil pH. E horizons a few feet thick in Florida usually have pH just above 5 (merely "strongly acid") while E horizons a few inches thick in New England are "extremely acid" with pH readings of 4.5 or below. In the southern Blue Ridge Mountains there are "ultra acid" soils, pH below 3.5, which have no E horizon.Presence of a caliche layer indicates the presence of calcium carbonates, which are present in alkaline conditions. Also, columnar structure can be an indicator of sodic condition.
- Observation of predominant flora. Calcifuge plants (those that prefer an acidic soil) include *Erica*, *Rhododendron* and nearly all other Ericaceae species, many birch (*Betula*), foxglove (*Digitalis*), gorse (*Ulex* spp.), and Scots Pine (*Pinus sylvestris*). Calcicole (lime loving) plants include ash trees (*Fraxinus* spp.),

honeysuckle (*Lonicera*), *Buddleja*, dogwoods (*Cornus* spp.), lilac (*Syringa*) and *Clematis* species.

- Use of an inexpensive pH testing kit, where in a small sample of soil is mixed with indicator solution which changes colour according to the acidity/alkalinity.
- Use of litmus paper. A small sample of soil is mixed with distilled water, into which a strip of litmus paper is inserted. If the soil is acidic the paper turns red, if alkaline, blue.
- Use of a commercially available electronic pH meter, in which a rod is inserted into moistened soil and measures the concentration of hydrogen ions.

Examples of Plant pH Preferences

- pH 4.5–5.0: Ericaceae (azalea, bilberry, blueberry, cranberry; heather), hydrangea for blue, (less acidic for pink), liquidambar or sweet gum, orchid, pin oak
- pH 5.0–5.5: Boronia, daphne; Ericaceae: (camellia, heatherrhododendron), ferns, iris, orchids, parsley, conifers (e.g., pine); Poaceae: (maize, millet, rye, oat), radish; Solanales: (potato, sweet potato); Bromeliaceae (pineapple).
- pH 5.5–6.0: Asteraceae: (aster, endive); Brassicaceae: (brussels sprout, kohlrabi), carrot; Cucurbitales: (begonia, chayote or choko); Fabaceae: (bean, crimson clover, peanut, soybean), petunia, rhubarb, violet, most bulbs (canna, daffodil, jonquil), larkspur primrose
- pH 6.0–6.5 antirrhinum or snapdragon, Brassicaceae: (broccoli, cabbage, candytuft, cauliflower, turnip, wallflower); Cucurbitaceae: (cucumber, pumpkin, squash); Fabaceae: (pea, red clover, white clover), gladiolus, Iceland poppy; Rosales: (cannabis, rose, strawberry); Solanaceae: (eggplant or aubergine, tomato), sweet corn; Violaceae: (pansy, viola), zinnia or zinnea
- pH 6.5–7.0: Amaranthaceae: (beet, spinach); Apiaceae: (celery, parsnip); Asparagales: (asparagus, onion); Asteraceae: (chrysanthemum, dahlia, lettuce), carnation; Fabaceae: (alfalfa, sweet pea), melon, stock tulip
- pH 7.1–8.0 lilac

Changing soil pH

Increasing pH of acidic soil

The most common amendment to increase soil pH is lime ($CaCO_3$ or $MgCO_3$), usually in the form of finely ground agricultural lime. The amount of lime needed to change pH is determined by the mesh size of the lime (how finely it is ground) and the buffering capacity of the soil. A high mesh size (60–100) indicates a finely ground lime, that will react quickly with soil acidity. Buffering capacity of soils is a function of a soils cation exchange capacity, which is in turn determined by the clay content of the soil, the type of clay and the amount of organic matter present. Soils with high clay content, particularly shrink–swell clay, will have a higher buffering capacity than soils with little clay. Soils with high organic matter will also have a higher buffering capacity than those with low organic matter. Soils with high buffering capacity require a greater amount of lime to be added than a soil with a lower buffering capacity for the same incremental change in pH.

Other amendments that can be used to increase the pH of soil include wood ash, industrial CaO (burnt lime), and oyster shells. White firewood ash includes metal salts which are important for processes requiring ions such as Na^+ (sodium), K^+ (potassium), Ca^{2+} (calcium), which may or may not be good for the select flora, but decreases the acidic quality of soil.

These products increase the pH of soils through the reaction of CO_3^{2-} with H^+ to produce CO_2 and H_2O. Calcium silicate neutralizes active acidity in the soil by removing free hydrogen ions, thereby increasing pH. As its silicate anion captures H^+ ions (raising the pH), it forms monosilicic acid (H_4SiO_4), a neutral solute.

Decreasing pH of Alkaline Soil

- Iron sulfates or aluminium sulfate as well as elemental sulfur (S) reduce pH through the formation of sulfuric acid.
- Urea, urea phosphate, ammonium nitrate, ammonium phosphates, ammonium sulfate and monopotassium phosphate fertilizers.
- organic matter in the form of plant litter, compost, and manure will decrease soil pH through the decomposition process. Certain acid organic matter such as pine needles, pine sawdust and acid peat are effective at reducing pH.

ACID SULFATE SOIL

Acid sulfate soils are naturally occurring soils, sediments or organic substrates (e.g. peat) that are formed under waterlogged conditions. These soils contain iron sulfide minerals (predominantly as the mineral pyrite) or their oxidation products. In an undisturbed state below the water table, acid sulfate soils are benign. However, if the soils are drained, excavated or exposed to air by a lowering of the water table, the sulfides react with oxygen to form sulfuric acid.

Release of this sulfuric acid from the soil can in turn release iron, aluminium, and other heavy metals (particularly arsenic) within the soil. Once mobilized in this way, the acid and metals can create a variety of adverse impacts: killing vegetation, seeping into and acidifying groundwaterand surface water bodies, killing fish and other aquatic organisms, and degrading concrete and steel structures to the point of failure.

Acid Sulfate Soil Formation

The soils and sediments most prone to becoming acid sulfate soils formed within the last 10,000 years, after the last major sea level rise. When the sea level rose and inundated the land, sulfate in the seawater mixed with land sediments containing iron oxides and organic matter. Under these anaerobic conditions, lithotrophic bacteria such as *Desulfovibrio desulfuricans* obtain oxygen for respiration through the reduction of sulfate ions in sea or groundwater, producing hydrogen sulfide. This in turn reacts with dissolved ferrous iron, forming very fine grained and highly reactive framboid crystals of iron sulfides such as (pyrite). Up to a point, warmer temperatures are more favourable conditions for these bacteria, creating a greater potential for formation of iron sulfides. Tropical waterlogged environments, such as mangrove swamps or estuaries, may contain higher levels of pyrite than those formed in more temperate climates.

Polders with acid sulfate soils in Guinea Bissau along a sea-arm amidst mangroves.

The pyrite is stable until exposed to air, at which point the pyrite rapidly oxidises and produces sulfuric acid. The impacts of acid sulfate soil leachate may persist over a long time, and/or peak seasonally (after dry periods with the first rains). In some areas of Australia, acid sulfate soils that drained 100 years ago are still releasing acid.

Chemical Reaction

When drained, pyrite (FeS_2) containing soils (also called cat-clays) may become extremely acidic ($pH < 4$) due to the oxidation of pyrite into sulfuric acid (H_2SO_4). In its simplest form, this chemical reaction is as follows:

$$2\ FeS_2 + 9\ O_2 + 4\ H_2O \rightarrow 8\ H^+ + 4\ SO_4^{2-} + 2\ Fe(OH)_3\ \text{(solid)}$$

The product $Fe(OH)_3$, iron(III) hydroxide (orange), precipitates as a solid, insoluble mineral by which the alkalinity component is immobilized, while the acidity remains active in the sulfuric acid. The process of acidification is accompanied by the formation of high amounts of aluminium (Al^{3+}, released from clay minerals under influence of the acidity), which are harmful to vegetation. Other products of the chemical reaction are:

- Hydrogen sulfide (H_2S), a smelly gas
- Sulfur (S), a yellow solid
- Iron(II) sulfide (FeS), a black/gray/blue solid
- Hematite (Fe_2O_3), a red solid

- Goethite (FeO.OH), a brown mineral
- Schwertmannite a brown mineral
- Iron sulfate compounds (e.g. jarosite)
- H-Clay (hydrogen clay, with a large fraction of adsorbed H^+ ions, a stable mineral, but poor in nutrients)

The iron can be present in bivalent and trivalent forms (Fe^{2+}, the ferrous ion, and Fe^{3+}, the ferric ion respectively). The ferrous form is soluble, whereas the ferric form is not. The more oxidized the soil becomes, the more the ferric forms dominate. Acid sulfate soils exhibit an array of colors ranging from black, brown, blue-gray, red, orange and yellow. The hydrogen clay can be improved by admitting sea water: the magnesium (Mg) and sodium (Na) in the sea water replaces the adsorbed hydrogen and other exchangeable acidic cations such as aluminium (Al). However this can create additional risks when the hydrogen ions and exchangeable metals are mobilised.

Geographical Distribution

Acid sulfate soils are widespread around coastal regions, and are also locally associated with freshwater wetlands and saline sulfate-rich groundwater in some agricultural areas. In Australia, coastal acid sulfate soils occupy an estimated 58,000 km^2, underlying coastal estuaries and floodplains near where the majority of the Australian population lives. Acid sulfate soil disturbance is often associated with dredging, excavation dewatering activities during canal, housing and marina developments. Droughts can also result in acid sulfate soil exposure and acidification.

Acid sulfate soils that have not been disturbed are called *potential acid sulfate soils (PASS)*. Acid sulfate soils that *have* been disturbed are called *actual acid sulfate soils (AASS)*.

Impact of Acid Sulfate Soil

Disturbing potential acid sulfate soils can have a destructive effect on plant and fish life, and on aquatic ecosystems. Flushing of acidic leachate to groundwater and surface waters can cause a number of impacts, including:

- Ecological damage to aquatic and riparian ecosystems through fish kills, increased fish disease outbreaks, dominance of acid-tolerant species, precipitation of iron, etc.
- Effects on estuarine fisheries and aquaculture projects (increased disease, loss of spawning area, etc.).
- Contamination of groundwater and surface water with arsenic, aluminium and other metals.
- Reduction in agricultural productivity through metal contamination of soils (predominantly by aluminium).

- Damage to infrastructure through the corrosion of concrete and steel pipes, bridges and other sub-surface assets.

Agricultural Impacts

Sea water is admitted to a bunded polder on acid sulfate soil for soil improvement and weed control, Guinea Bissau

Potentially acid sulfate soils (also called cat-clays) are often not cultivated or, if they are, planted with rice, so that the soil can be kept wet preventing oxidation. Subsurface drainage of these soils is normally not advisable.

When cultivated, acid sulfate soils cannot be kept wet continuously because of climatic dry spells and shortages of irrigation water, surface drainage may help to remove the acidic and toxic chemicals (formed in the dry spells) during rainy periods. In the long run surface drainage can help to reclaim acid sulfate soils. The indigenous population of Guinea Bissau has thus managed to develop the soils, but it has taken them many years of careful management and toil.

In an article on cautious land drainage, the author describes the successful application of subsurface drainage in acid sulfate soils in coastal polders of Kerala state, India.

Also in the Sunderbans, West Bengal, India, acid sulfate soils have been taken in agricultural use.

A study in South Kalimantan, Indonesia, in a perhumid climate, has shown that the acid sulfate soils with a widely spaced subsurface drainage system have yielded promising results for the cultivation of upland (sic!) rice, peanut and soybean. The local population, of old, had already settled in this area and were able to produce a variety of crops (including tree fruits), using hand-dug drains running from the river into the land until reaching the back swamps. The crop yields were modest, but provided enough income to make a decent living.

Reclaimed acid sulfate soils have a well-developed soil structure; they are well permeable, but infertile due to the leaching that has occurred.

In the second half of the 20th century, in many parts of the world, waterlogged and potentially acid sulfate soils have been drained aggressively to make them productive for agriculture. The results were disastrous. The soils are unproductive, the lands look barren and the water is very clear, devoid of silt and life. The soils can be colorful, though.

Construction

When brickwork is persistently wet, as in foundations, retaining walls, parapets and chimneys, sulfates in bricks and mortar may in time crystallise and expand and cause mortar and renderings to disintegrate. To minimise this effect specialised brickwork with low sulfate levels should be used. Acid sulfates that are located within the subsoil strata has the same effects on the foundations of a building. Adequate protection can exist using a polythene sheeting to encase the foundations or using a sulfate-resistant Portland cement. To identify the pH level of the ground a soil investigation must take place.

Acid Sulfate Soil Restoration and Management

By raising the water table, after damage has been inflicted due to over-intensive drainage, the soils can be restored. The following table gives an example.

Drainage and yield of Malaysian oil palm on acid sulfate soils (after Toh Peng Yin and Poon Yew Chin, 1982) Yield in tons of fresh fruit per ha:

Year	60	61	62	63	64	65	66	67	68	69	70	71
Yield	17	14	15	12	8	2	4	8	14	19	18	19

Drainage depth and intensity were increased in 1962. The water table was raised again in 1966 to counter negative effects.

In the "millennium drought" in the Murray-Darling Basin in Australia, exposure of acid sulfate soils occurred. Large scale engineering interventions were undertaken to prevent further acidification, including construction of a bund and pumping of water to prevent exposure and acidification of Lake Albert. Management of acidification in the Lower Lakes was also undertaken using aerial limestone dosing.

ESTUARINE ACIDIFICATION

Estuarine acidification is a decrease in the pH of coastal marine ecosystems, specifically those of estuaries. pH change in estuaries is more complicated than in the open ocean due to direct impacts from land run-off and coastal current dynamics. Ocean acidification is the ongoing decrease in the pH of the Earth's oceans, caused by the absorption of carbon dioxide (CO_2) from the atmosphere (0.1 units over the last century). The ocean absorbs 30-40% of all CO_2 emitted to the atmosphere; this increase in aqueous carbon dioxide causes a decline in the pH of the ocean surface. As carbon dioxide combines with water, it releases protons (hydrogen ions), based on the following equation:

$$CO_2 + H_2O \leftrightarrow H_2CO_3 \leftrightarrow HCO_3 + H^+ \leftrightarrow CO_3 + 2\ H^+$$

Causes of variable pH

Freshwater Flow

An estuary is defined as “a water passage where the tide meets a river current”. The pH of estuaries is highly variable because of freshwater flow from rivers and groundwater, as well as primary productivity (exacerbated by nutrient loading) and coastal upwelling. Fresh water from rivers typically has a lower pH than ocean water (~7 compared to ~8). Seasonal and annual changes in river flow entering an estuary can change the pH by whole units.

Photosynthesis and Respiration

Primary production (plant growth) changes pH on a daily, seasonal, and annual basis. During photosynthesis, carbon dioxide is removed from the water, increasing pH. Organisms release carbon dioxide during respiration. This leads to a daily cycle of increased pH during daylight hours and a decrease in pH during the night, when respiration is dominant. Similarly, pH is higher during the winter when grazing is low compared to productivity.

Effluent

Many estuaries experience nutrient loading from runoff containing wastewater effluent or fertilizers, natural or artificial. Increased nutrients can stimulate primary productivity and alter the balance between primary productivity and respiration. This process can change pH by whole units within the estuary. Both these processes make it difficult to measure the overall change in pH associated with increased atmospheric carbon dioxide levels.

Currents

Areas with coastal upwelling such as the west coast of North America have experienced increases in acidification due to more acidic deep water upwelling into the estuary. This may have a detrimental effect on the survival of calcifying organisms because the organisms have a much more difficult time forming and maintaining their calcium carbonate shells.

Impact on Marine Life

A coccolithophore with many coccoliths (plates) formed from calcium carbonate.

As the pH of marine systems decreases, it causes calcium carbonate ($CaCO_3$) to dissociate to keep in chemical equilibrium. Calcium carbonate is vital to calcifying organisms such as shellfish, corals, and coccolithophores (a type of phytoplankton). Acidification also harms micro-organisms in the environment. These organisms either directly provide humans with a food source or supports an ecosystem important to humans.

Research

Estuarine acidification is being intensively studied to understand the biological, chemical, and physical factors that affect pH in estuaries.

REFERENCES

1. Perry, Leonard. "pH for the Garden". Retrieved 11 December 2012.
2. Soil Survey Division Staff. "Soil survey manual.1993. Chapter 3, selected chemical properties.". Soil Conservation Service. U.S. Department of Agriculture Handbook 18. Retrieved 2011-03-12.
3. Sparks, Donald; Environmental Soil Chemistry. 2003, Academic Press, London, UK
4. http://edis.ifas.ufl.edu/ch086
5. Finck, Arnold (1976). Pflanzenernährung in Stichworten. Kiel: Hirt. p. 80. ISBN 3-554-80197-6.
6. Brady, N. and Weil, R. The Nature and Properties of Soils. 13th ed. 2002
7. Hansson et al (2011) Differences in soil properties in adjacent stands of Scots pine, Norway spruce and silver birch in SW Sweden. Forest Ecology and Management 262 522–530

8. Rout, GR; Samantaray, S; Das, P (2001). "Aluminium toxicity in plants: a review" (PDF). Agronomie. 21 (1): 4–5. doi:10.1051/agro:2001105. Retrieved 11 June 2014.
9. http://www.extension.org/pages/9875/soil-ph-and-nutrient-availability
10. Soilinfo PSU
11. USDA Myakka data
12. USDA Berkshire data
13. USDA Cataloochee data
14. Buol, S. W., R. J. Southard, R.C. Graham and P.A. McDaniel. Soil Genesis and Classification. (5th) Edition, Ia. State Press p. 494. 2002
15. http://www2.hawaii.edu/~nvhue/acid.html
16. Brady, N. and Weil, R. The Nature and Properites of Soils. 13th ed. 2002

CHAPTER 5

Soil Bacteria

Chapter Contents

- **Introduction**
- **Etymology**
- **Origin and Early Evolution**
- **Morphology**
- **Cellular Structure**
- **Metabolism**
- **Growth and Reproduction**
- **Genomes**
- **Genetics**
- **Behaviour**
- **Classification and Identification**
- **Interactions with Other Organisms**

Bacteria are single celled microbes. The cell structure is simpler than that of other organisms as there is no nucleus or membrane bound organelles. Instead their control center containing the genetic information is contained in a single loop of DNA. Some bacteria have an extra circle of genetic material called a plasmid.

INTRODUCTION

Bacteria constitute a large domain of prokaryotic microorganisms. Typically a few micrometres in length, bacteria have a number of shapes, ranging from spheres to rods and spirals. Bacteria were among the first life forms to appear on Earth, and are present in most of its habitats. Bacteria inhabit soil, water, acidic hot springs, radioactive waste, and the deep portions of Earth's crust. Bacteria also live in symbiotic and parasitic relationships with plants and animals.

There are typically 40 million bacterial cells in a gram of soil and a million bacterial cells in a millilitre of fresh water. There are approximately 5×10^{30} bacteria on Earth, forming a biomass which exceeds that of all plants and animals. Bacteria are vital in recycling nutrients, with many of the stages in nutrient cycles dependent on these organisms, such as the fixation of nitrogen from the atmosphere and putrefaction. In the biological communities surrounding hydrothermal vents and cold seeps, bacteria provide the nutrients needed to sustain life by converting dissolved compounds, such as hydrogen sulphide and methane, to energy. On 17 March 2013, researchers reported data that suggested bacterial life forms thrive in the Mariana Trench, which with a depth of up to 11 kilometres is the deepest part of the Earth's oceans.Other researchers reported related studies that microbes thrive inside rocks up to 580 metres below the sea floor under 2.6 kilometres of ocean off the coast of the northwestern United States. According to one of the researchers, You can find microbes everywhere — they're extremely adaptable to conditions, and survive wherever they are.

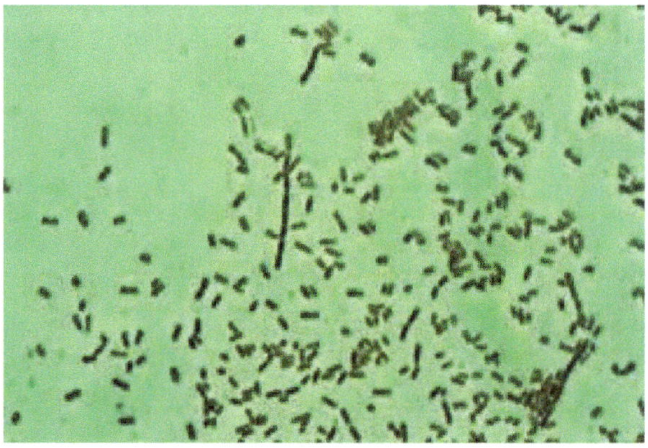

Most bacteria have not been characterised, and only about half of the bacterial phyla have species that can be grown in the laboratory. The study of bacteria is known as bacteriology, a branch of microbiology.

There are approximately ten times as many bacterial cells in the human flora as there are human cells in the body, with the largest number of the human flora being in the gut flora, and a large number on the skin. The vast majority of the bacteria in the body are rendered harmless by the protective effects of the immune system, and some are beneficial. However, several species of bacteria are pathogenic and cause infectious diseases, including cholera, syphilis, anthrax, leprosy, and bubonic plague. The most common fatal bacterial diseases are respiratory infections, with tuberculosis alone killing about 2 million people per year, mostly in sub-Saharan Africa. In developed countries, antibiotics are used to treat bacterial infections and are also used in farming, making antibiotic resistance a growing problem. In industry, bacteria are important in sewage treatment and the breakdown of oil spills, the production of cheese and yogurt through fermentation, and the recovery of gold, palladium, copper and other metals in the mining sector, as well as in biotechnology, and the manufacture of antibiotics and other chemicals.

Once regarded as plants constituting the class *Schizomycetes*, bacteria are now classified as prokaryotes. Unlike cells of animals and other eukaryotes, bacterial cells do not contain a nucleus and rarely harbour membrane-bound organelles. Although the term *bacteria* traditionally included all prokaryotes, the scientific classification changed after the discovery in the 1990s that prokaryotes consist of two very different groups of organisms that evolved from an ancient common ancestor. These evolutionary domains are called *Bacteria* and *Archaea*.

ETYMOLOGY

The word *bacteria* is the plural of the New Latin *bacterium*, which is the latinisation of the Greekβακτήριον (*bakterion*), the diminutive of βακτηρία (*bakteria*), meaning "staff, cane",because the first ones to be discovered were rod-shaped.

ORIGIN AND EARLY EVOLUTION

The ancestors of modern bacteria were unicellular microorganisms that were the first forms of life to appear on Earth, about 4 billion years ago. For about 3 billion years, most organisms were microscopic, and bacteria and archaea were the dominant forms of life. In 2008, fossils of macroorganisms were discovered and named as the Francevillian biota. Although bacterial fossils exist, such as stromatolites, their lack of distinctive morphology prevents them from being used to examine the history of bacterial evolution, or to date the time of origin of a particular bacterial species. However, gene sequences can be used to reconstruct the bacterial phylogeny, and these studies indicate that bacteria diverged first from the archaeal/eukaryotic lineage. Bacteria were also involved in the second great evolutionary divergence, that of the archaea and eukaryotes. Here, eukaryotes resulted from the entering of ancient bacteria into endosymbiotic associations with the ancestors of eukaryotic cells, which were themselves possibly related to the Archaea. This involved the engulfment by proto-eukaryotic cells of alphaproteobacterial symbionts to form either mitochondria or hydrogenosomes, which are still found in all known Eukarya (sometimes in highly reduced form, e.g. in ancient amitochondrial protozoa). Later on, some eukaryotes that already contained mitochondria also engulfed cyanobacterial-like organisms. This led to the formation of chloroplasts in algae and plants. There are also some algae that originated from even later endosymbiotic events. Here, eukaryotes engulfed a eukaryotic algae that developed into a "second-generation" plastid.This is known as secondary endosymbiosis.

MORPHOLOGY

Bacteria display a wide diversity of shapes and sizes, called *morphologies*. Bacterial cells are about one-tenth the size of eukaryotic cells and are typically 0.5–5.0 micrometres in length. However, a few species are visible to the unaided eye — for example, *Thiomargarita namibiensis* is up to half a millimetre long and *Epulopiscium fishelsoni* reaches 0.7 mm. Among the smallest bacteria are members of the genus *Mycoplasma*, which measure only 0.3 micrometres, as small as the largest viruses.Some bacteria may be even smaller, but these ultramicrobacteria are not well-studied.

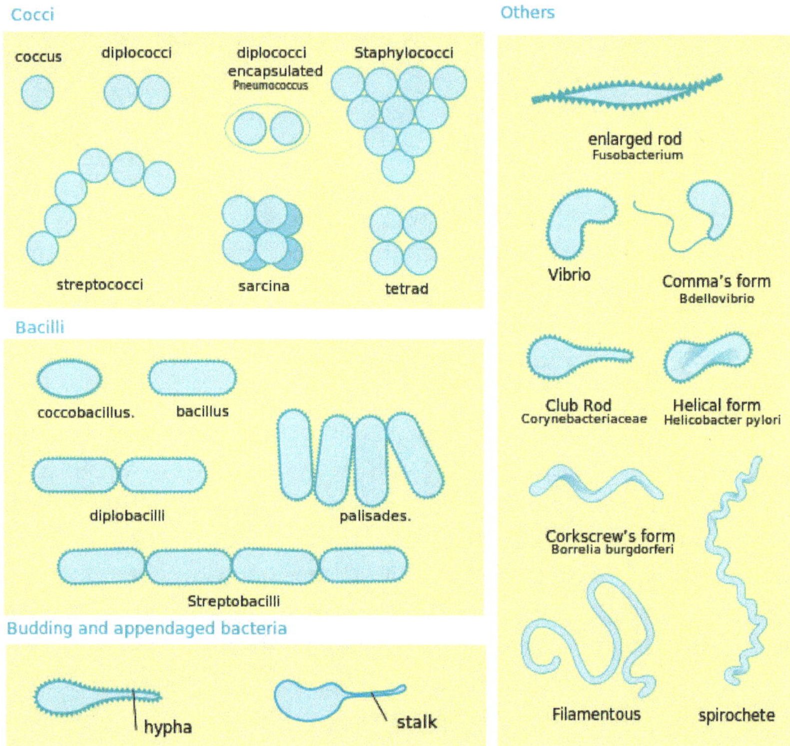

Bacteria display many cell morphologies and arrangements

Most bacterial species are either spherical, called *cocci* (*sing*. coccus, from Greek *kókkos*, grain, seed), or rod-shaped, called *bacilli* (*sing*. bacillus, from Latin *baculus*, stick). Elongation is associated with swimming. Some bacteria, called *vibrio*, are shaped like slightly curved rods or comma-shaped; others can be spiral-shaped, called *spirilla*, or tightly coiled, called *spirochaetes*. A small number of species even have tetrahedral or cuboidal shapes. More recently, some bacteria were discovered deep under Earth's crust that grow as branching filamentous types with a star-shaped cross-section. The large surface area to volume ratio of this morphology may give these bacteria an advantage in nutrient-poor environments. This wide variety of shapes is determined by the bacterial cell wall and cytoskeleton, and is important because it can influence the ability of bacteria to acquire nutrients, attach to surfaces, swim through liquids and escape predators.

Many bacterial species exist simply as single cells, others associate in characteristic patterns: *Neisseria* form diploids (pairs), *Streptococcus* form chains, and *Staphylococcus* group together in "bunch of grapes" clusters. Bacteria can also be elongated to form filaments, for example the Actinobacteria. Filamentous bacteria are often surrounded by a sheath that contains many individual cells. Certain types, such as species of the genus *Nocardia*, even form complex, branched filaments, similar in appearance to fungal mycelia.

A biofilm of thermophilic bacteria in the outflow of Mickey Hot Springs, Oregon, approximately 20 mm thick.

Bacteria often attach to surfaces and form dense aggregations called *biofilms* or bacterial mats. These films can range from a few micrometres in thickness to up to half a metre in depth, and may contain multiple species of bacteria, protists and archaea. Bacteria living in biofilms display a complex arrangement of cells and extracellular components, forming secondary structures, such as microcolonies, through which there are networks of channels to enable better diffusion of nutrients. In natural environments, such as soil or the surfaces of plants, the majority of bacteria are bound to surfaces in biofilms. Biofilms are also important in medicine, as these structures are often present during chronic bacterial infections or in infections of implanted medical devices, and bacteria protected within biofilms are much harder to kill than individual isolated bacteria.

Even more complex morphological changes are sometimes possible. For example, when starved of amino acids, Myxobacteria detect surrounding cells in a process known as quorum sensing, migrate towards each other, and aggregate to form fruiting bodies up to 500 micrometres long and containing approximately 100,000 bacterial cells. In these fruiting bodies, the bacteria perform separate tasks; this type of cooperation is a simple type of multicellular organisation. For example, about one in 10 cells migrate to the top of these fruiting bodies and differentiate into a specialised dormant state called myxospores, which are more resistant to drying and other adverse environmental conditions than are ordinary cells.

CELLULAR STRUCTURE

Intracellular Structures

The bacterial cell is surrounded by a cell membrane (also known as a lipid, cytoplasmic or plasma membrane). This membrane encloses the contents of the cell and acts as a barrier to hold nutrients, proteins and other essential components of the *cytoplasm* within the cell. As they are prokaryotes, bacteria do not usually have membrane-bound organelles in their cytoplasm, and thus contain few large intracellular structures. They lack a true nucleus, mitochondria, chloroplasts and the other organelles present in eukaryotic cells. Bacteria were once seen as simple bags of cytoplasm, but structures such as the *prokaryotic cytoskeleton* and the localisation of proteins to specific locations within the cytoplasm that give bacteria some complexity have been discovered. These subcellular levels of organisation have been called bacterial hyperstructures.

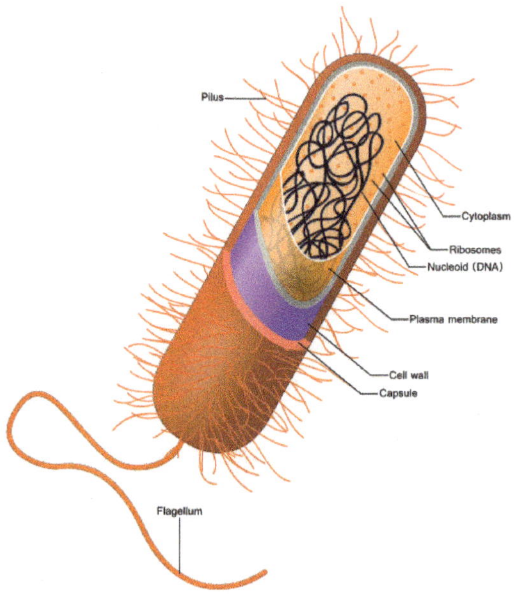

Structure and contents of a typical gram-positive bacterial cell (seen by the fact that only *one* cell membrane is present).

Bacterial microcompartments, such as carboxysomes, provide a further level of organisation; they are compartments within bacteria that are surrounded by polyhedral protein shells, rather than by lipid membranes. These "polyhedral organelles" localise and compartmentalise bacterial metabolism, a function performed by the membrane-bound organelles in eukaryotes.

Many important biochemical reactions, such as energy generation, use concentration gradients across membranes. The general lack of internal membranes in bacteria means reactions such as electron transport occur across the cell membrane between the cytoplasm and the *periplasmic space*. However, in many photosynthetic bacteria the plasma membrane is highly folded and fills most of the cell with layers of light-gathering membrane. These light-gathering complexes may even form lipid-enclosed structures called chlorosomes in green sulfur bacteria. Other proteins import nutrients across the cell membrane, or expel undesired molecules from the cytoplasm.

Bacteria do not have a membrane-bound nucleus, and their genetic material is typically a single circular DNA chromosome located in the cytoplasm in an irregularly shaped body called the *nucleoid*. The nucleoid contains the chromosome with its associated proteins and RNA. The phylum Planctomycetes and candidate phylum Poribacteria may be exceptions to the general absence of internal membranes in bacteria, because they appear to have a double membrane around their nucleoids and contain other membrane-bound cellular structures. Like all living organisms, bacteria contain *ribosomes*, often grouped in chains called polyribosomes, for the production of proteins, but the structure of the bacterial ribosome is different from that of eukaryotes and Archaea. Bacterial ribosomes have a sedimentation rate of 70S (measured in Svedberg units): their subunits have rates of 30S and 50S. Some antibiotics bind specifically to 70S ribosomes and inhibit bacterial protein synthesis. Those antibiotics kill bacteria without affecting the larger 80S ribosomes of eukaryotic cells and without harming the host.

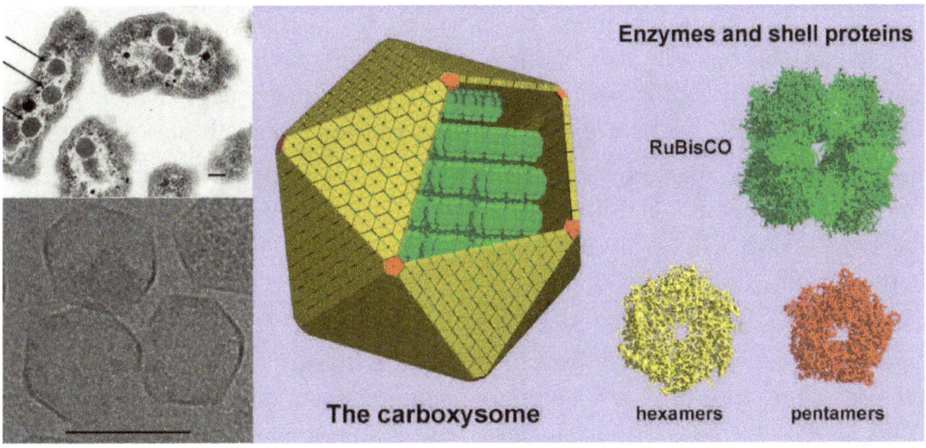

Carboxysomes are protein-enclosed bacterial organelles. Top left is an electron microscope image of carboxysomes in *Halothiobacillus neapolitanus*, below is an image of purified carboxysomes. On the right is a model of their structure. Scale bars are 100 nm.

Some bacteria produce intracellular nutrient storage granules for later use, such as glycogen, polyphosphate, sulfur or polyhydroxyalkanoates. Certain bacterial species, such as the photosynthetic Cyanobacteria, produce internal gas vesicles, which they use to regulate their buoyancy – allowing them to move up or down into water layers with different light intensities and nutrient levels. *Intracellular membranes* called *chromatophores* are also found in membranes of phototrophic bacteria. Used primarily for photosynthesis, they contain bacteriochlorophyll pigments and carotenoids. An early idea was that bacteria might contain membrane folds termed mesosomes, but these were later shown to be artefacts produced by the chemicals used to prepare the cells for electron microscopy. *Inclusions* are considered to be nonliving components of the cell that do not possess metabolic activity and are not bounded by membranes. The most common inclusions are glycogen, lipid droplets, crystals, and pigments. *Volutin granules* are cytoplasmic inclusions of complexed inorganic polyphosphate. These granules are called *metachromatic granules* due to their displaying the metachromatic effect; they appear red or blue when stained with the blue dyes methylene blue or toluidine blue. *Gas vacuoles*, which are freely permeable to gas, are membrane-bound vesicles present in some species of *Cyanobacteria*. They allow the bacteria to control their buoyancy. *Microcompartments* are widespread, membrane-bound organelles that are made of a protein shell that surrounds and encloses various enzymes. *Carboxysomes* are bacterial microcompartments that contain enzymes involved in carbon fixation. *Magnetosomes* are bacterial microcompartments, present in magnetotactic bacteria, that contain magnetic crystals.

Extracellular Structures

In most bacteria, a *cell wall* is present on the outside of the cell membrane. The cell membrane and cell wall comprise the *cell envelope*. A common bacterial cell wall material is *peptidoglycan* (called "murein in older sources), which is made from polysaccharide chains cross-linked by peptides containing D-amino acids. Bacterial cell walls are different from the cell walls of plants and fungi, which are made of cellulose and chitin, respectively. The cell wall of bacteria is also distinct from that of Archaea, which do not contain peptidoglycan. The cell wall is essential to the survival of many bacteria, and the antibiotic penicillin is able to kill bacteria by inhibiting a step in the synthesis of peptidoglycan.

There are broadly speaking two different types of cell wall in bacteria, a thick one in the gram-positives and a thinner one in the gram-negatives. The names originate from the reaction of cells to the Gram stain, a long-standing test for the classification of bacterial species.

Gram-positive bacteria possess a thick cell wall containing many layers of peptidoglycan and *teichoic acids*. In contrast, *gram-negative bacteria* have a relatively thin cell wall consisting of a few layers of peptidoglycan surrounded by a second lipid membrane containing *lipopolysaccharides* and lipoproteins. Lipopolysaccharides, also called *endotoxins*, are composed of polysaccharides and *lipid A* that is responsible for much of the toxicity of gram-negative bacteria. Most bacteria have the gram-negative cell wall, and only the Firmicutes and Actinobacteria have the alternative gram-positive arrangement. These two groups were previously known as the low G+C and high G+C gram-positive bacteria, respectively. These differences in structure can produce differences in antibiotic susceptibility; for instance, vancomycin can kill only gram-positive bacteria and is ineffective against gram-negative pathogens, such as *Haemophilus influenzae* or *Pseudomonas aeruginosa*. If the bacterial cell wall is entirely removed, it is called a *protoplast*, whereas if it is partially removed, it is called a *spheroplast*. β-Lactam antibiotics, such as penicillin, inhibit the formation of peptidoglycan cross-links in the bacterial cell wall. The enzyme lysozyme, found in human tears, also digests the cell wall of bacteria and is the body's main defence against eye infections.

Acid-fast bacteria, such as *Mycobacteria*, are resistant to decolorisation by acids during staining procedures. The high mycolic acid content of *Mycobacteria*, is responsible for the staining pattern of poor absorption followed by high retention. The most common staining technique used to identify acid-fast bacteria is the Ziehl-Neelsen stain or acid-fast stain, in which the acid-fast bacilli are stained bright-red and stand out clearly against a blue background. *L form bacteria* are strains of bacteria that lack cell walls. The main pathogenic bacteria in this class is *Mycoplasma* (not to be confused with *Mycobacteria*).

In many bacteria, an *S-layer* of rigidly arrayed protein molecules covers the outside of the cell. This layer provides chemical and physical protection for the cell surface and can act as a macromolecular diffusion barrier. S-layers have diverse but mostly poorly understood functions, but are known to act as virulence factors in *Campylobacter* and contain surface enzymes in *Bacillus stearothermophilus*.

Flagella are rigid protein structures, about 20 nanometres in diameter and up to 20 micrometres in length, that are used for motility. Flagella are driven by the energy released by the transfer of ions down an electrochemical gradient across the cell membrane.

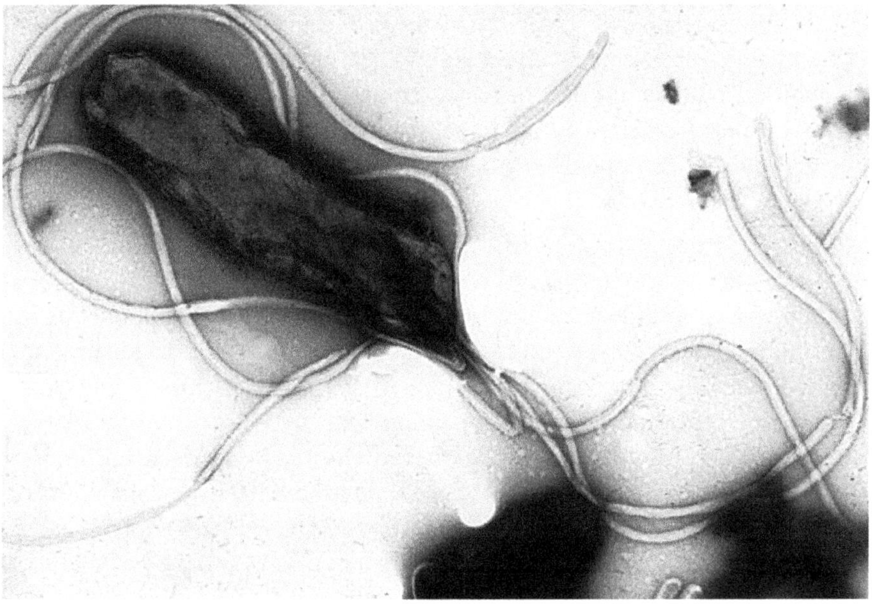

Helicobacter pylori electron micrograph, showing multiple flagella on the cell surface

Fimbriae (sometimes called "attachment pili) are fine filaments of protein, usually 2–10 nanometres in diameter and up to several micrometres in length. They are distributed over the surface of the cell, and resemble fine hairs when seen under the electron microscope. Fimbriae are believed to be involved in attachment to solid surfaces or to other cells, and are essential for the virulence of some bacterial pathogens. *Pili* (*sing.* pilus) are cellular appendages, slightly larger than fimbriae, that can transfer genetic material between bacterial cells in a process called conjugation where they are called *conjugation pili* or "sex pili (see bacterial genetics, below). They can also generate movement where they are called *type IV pili* (see movement, below).

Glycocalyx are produced by many bacteria to surround their cells, and vary in structural complexity: ranging from a disorganised *slime layer* of extra-cellular polymer to a highly structured *capsule*. These structures can protect cells from engulfment by eukaryotic cells such as macrophages (part of the human immune system). They can also act as antigens and be involved in cell recognition, as well as aiding attachment to surfaces and the formation of biofilms.

The assembly of these extracellular structures is dependent on bacterial secretion systems. These transfer proteins from the cytoplasm into the periplasm or into the environment around the cell. Many types of secretion systems are known and these structures are often essential for the virulence of pathogens, so are intensively studied.

Endospores

Certain genera of gram-positive bacteria, such as *Bacillus*, *Clostridium*, *Sporohalobacter*, *Anaerobacter*, and *Heliobacterium*, can form highly resistant, dormant structures called *endospores*. In almost all cases, one endospore is formed and this is not a reproductive process, although *Anaerobacter* can make up to seven endospores in a single cell. Endospores have a central core of cytoplasm containing DNA and ribosomes surrounded by a cortex layer and protected by an impermeable and rigid coat. Dipicolinic acid is a chemical compound that composes 5% to 15% of the dry weight of bacterial spores. It is implicated as responsible for the heat resistance of the endospore.

Endospores show no detectable metabolism and can survive extreme physical and chemical stresses, such as high levels of UV light, gamma radiation, detergents, disinfectants, heat, freezing, pressure, and desiccation. In this dormant state, these organisms may remain viable for millions of years, and endospores even allow bacteria to survive exposure to the vacuum and radiation in space. According to scientist Dr. Steinn Sigurdsson, There are viable bacterial spores that have been found that are 40 million years old on Earth — and we know they're very hardened to radiation.Endospore-forming bacteria can also cause disease: for example, anthrax can be contracted by the inhalation of *Bacillus anthracis* endospores, and contamination of deep puncture wounds with *Clostridium tetani* endospores causes tetanus.

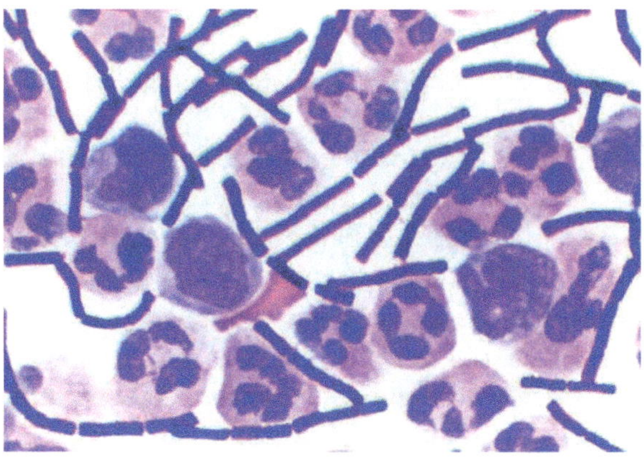

Bacillus anthracis (stained purple) growing in cerebrospinal fluid

METABOLISM

Bacteria exhibit an extremely wide variety of metabolic types. The distribution of metabolic traits within a group of bacteria has traditionally been used to define their taxonomy, but these traits often do not correspond with modern genetic classifications. Bacterial metabolism is classified into nutritional groups on the basis of three major criteria: the kind of energy used for growth, the source of carbon, and the electron donors used for growth. An additional criterion of respiratory microorganisms are the electron acceptors used for aerobic or anaerobic respiration.

Nutritional types in bacterial metabolism

Nutritional type	Source of energy	Source of carbon	Examples
Phototrophs	Sunlight	Organic compounds (photoheterotrophs) or carbon fixation (photoautotrophs)	Cyanobacteria, Green sulfur bacteria, Chloroflexi, or Purple bacteria
Lithotrophs	Inorganic compounds	Organic compounds (lithoheterotrophs) or carbon fixation (lithoautotrophs)	Thermodesulfobacteria, *Hydrogenophilaceae*, or Nitrospirae
Organotrophs	Organic compounds	Organic compounds (chemoheterotrophs) or carbon fixation (chemoautotrophs)	*Bacillus*, *Clostridium* or *Enterobacteriaceae*

Carbon metabolism in bacteria is either *heterotrophic*, where organic carbon compounds are used as carbon sources, or *autotrophic*, meaning that cellular carbon is obtained by fixing carbon dioxide. Heterotrophic bacteria include parasitic types. Typical autotrophic bacteria are phototrophic cyanobacteria, green sulfur-bacteria and some purple bacteria, but also many chemolithotrophic species, such as nitrifying or sulfur-oxidising bacteria. Energy metabolism of bacteria is either based on *phototrophy*, the use of light through photosynthesis, or based on *chemotrophy*, the use of chemical substances for energy, which are mostly oxidised at the expense of oxygen or alternative electron acceptors (aerobic/anaerobic respiration).

Bacteria are further divided into *lithotrophs* that use inorganic electron donors and *organotrophs* that use organic compounds as electron donors. Chemotrophic organisms use the respective electron donors for energy conservation (by aerobic/anaerobic respiration or fermentation) and biosynthetic reactions (e.g., carbon dioxide fixation), whereas phototrophic organisms use them only for biosynthetic purposes. Respiratory organisms use chemical compounds as a source of energy by taking electrons from the reduced substrate and transferring them to a terminal electron acceptor in a redox reaction. This reaction releases energy that can be used to synthesise ATP and drive metabolism. In *aerobic organisms*, oxygen is used as the electron acceptor. In *anaerobic organisms* other inorganic compounds, such as nitrate, sulfate or carbon dioxide are used as electron acceptors. This leads to the ecologically important processes of denitrification, sulfate reduction, and acetogenesis, respectively.

Another way of life of chemotrophs in the absence of possible electron acceptors is fermentation, wherein the electrons taken from the reduced substrates are transferred to oxidised intermediates to generate reduced fermentation products (e.g., lactate, ethanol, hydrogen, butyric acid). Fermentation is possible, because the energy content of the substrates is higher than that of the products, which allows the organisms to synthesise ATP and drive their metabolism.

These processes are also important in biological responses to pollution; for example, sulfate-reducing bacteria are largely responsible for the production of the highly toxic forms of mercury (methyl- and dimethylmercury) in the environment. Non-respiratory anaerobes use fermentation to generate energy and reducing power, secreting metabolic by-products (such as ethanol in brewing) as waste. Facultative anaerobes can switch between fermentation and different terminal electron acceptors depending on the environmental conditions in which they find themselves.

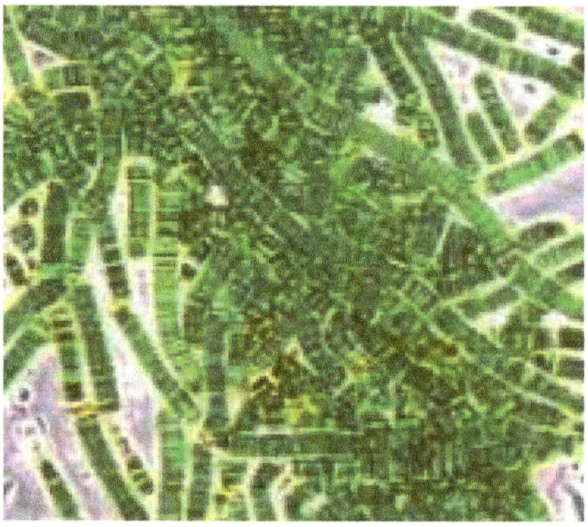

Filaments of photosyntheticcyanobacteria

Lithotrophic bacteria can use inorganic compounds as a source of energy. Common inorganic electron donors are hydrogen, carbon monoxide, ammonia (leading to nitrification), ferrous iron and other reduced metal ions, and several reduced sulfur compounds. In unusual circumstances, the gas methane can be used by methanotrophic bacteria as both a source of electrons and a substrate for carbon anabolism. In both aerobic phototrophy and chemolithotrophy, oxygen is used as a terminal electron acceptor, whereas under anaerobic conditions inorganic compounds are used instead. Most lithotrophic organisms are autotrophic, whereas organotrophic organisms are heterotrophic.

In addition to fixing carbon dioxide in photosynthesis, some bacteria also fix nitrogen gas (nitrogen fixation) using the enzyme nitrogenase. This environmentally important trait can be found in bacteria of nearly all the metabolic types listed above, but is not universal.

Regardless of the type of metabolic process they employ, the majority of bacteria are able to take in raw materials only in the form of relatively small molecules, which enter the cell by diffusion or through molecular channels in cell membranes. The Planctomycetes are the exception (as they are in possessing membranes around their nuclear material). It has recently been shown that *Gemmata obscuriglobus* is able to take in large molecules via a process that in some ways resembles endocytosis, the process used by eukaryotic cells to engulf external items.

GROWTH AND REPRODUCTION

Unlike in multicellular organisms, increases in cell size (cell growth) and reproduction by cell division are tightly linked in unicellular organisms. Bacteria grow to a fixed size and then reproduce through *binary fission*, a form of asexual reproduction. Under optimal conditions, bacteria can grow and divide extremely rapidly, and bacterial populations can double as quickly as every 9.8 minutes. In cell division, two identical clone daughter cells are produced. Some bacteria, while still reproducing asexually, form more complex reproductive structures that help disperse the newly formed daughter cells. Examples include fruiting body formation by *Myxobacteria* and aerial hyphae formation by *Streptomyces*, or budding. Budding involves a cell forming a protrusion that breaks away and produces a daughter cell.

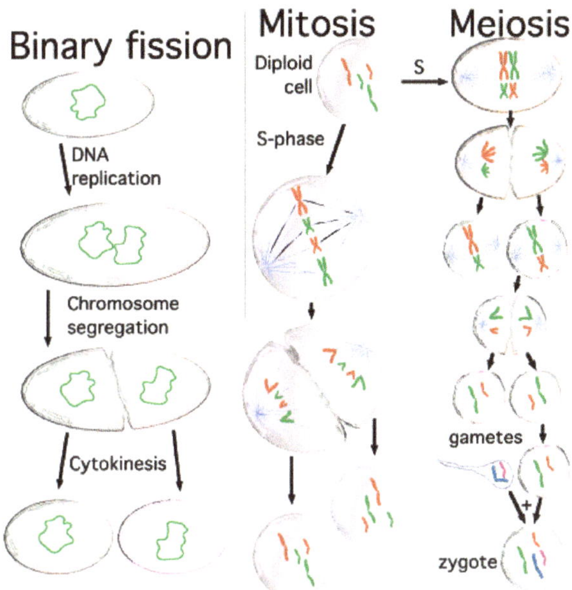

Many bacteria reproduce through binary fission, which is compared to mitosis and meiosis in this image.

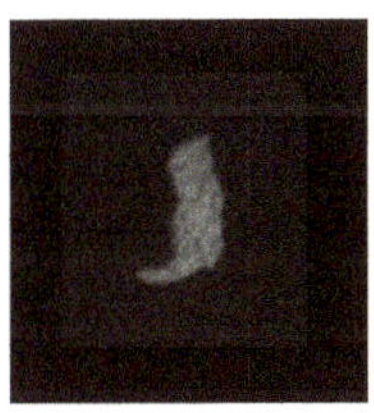

A colony of *Escherichia coli*

In the laboratory, bacteria are usually grown using solid or liquid media. Solid *growth media*, such as agar plates, are used to isolate pure cultures of a bacterial strain. However, liquid growth media are used when measurement of growth or large volumes of cells are required. Growth in stirred liquid media occurs as an even cell suspension, making the cultures easy to divide and transfer, although isolating single bacteria from liquid media is difficult. The use of selective media (media with specific nutrients added or deficient, or with antibiotics added) can help identify specific organisms.

Most laboratory techniques for growing bacteria use high levels of nutrients to produce large amounts of cells cheaply and quickly. However, in natural environments, nutrients are limited, meaning that bacteria cannot continue to reproduce indefinitely. This nutrient limitation has led the evolution of different growth strategies (see r/K selection theory). Some organisms can grow extremely rapidly when nutrients become available, such as the formation of algal (and cyanobacterial) blooms that often occur in lakes during the summer. Other organisms have adaptations to harsh environments, such as the production of multiple antibiotics by *Streptomyces* that inhibit the growth of competing microorganisms. In nature, many organisms live in communities (e.g., biofilms) that may allow for increased supply of nutrients and protection from environmental stresses. These relationships can be essential for growth of a particular organism or group of organisms (syntrophy).

Bacterial growth follows four phases. When a population of bacteria first enter a high-nutrient environment that allows growth, the cells need to adapt to their new environment. The first phase of growth is the *lag phase*, a period of slow growth when the cells are adapting to the high-nutrient environment and preparing for fast growth. The lag phase has high biosynthesis rates, as proteins necessary for rapid growth are produced.The second phase of growth is the *log phase*, also known as the *logarithmic or exponential phase*. The log phase is marked by rapid exponential growth. The rate at which cells grow during this phase is known as the *growth rate* (*k*), and the time it takes the cells to double is known as the *generation time* (*g*). During log phase, nutrients are metabolised at maximum speed until one of the nutrients is depleted and starts limiting growth. The third phase of growth is the *stationary phase* and is caused by depleted nutrients. The cells reduce their metabolic activity and consume non-essential cellular proteins. The stationary phase is a transition from rapid growth to a stress response state and there is increased expression of genes involved in DNA repair, antioxidant metabolism and nutrient transport. The final phase is the *death phase* where the bacteria run out of nutrients and die.

GENOMES

The genomes of thousands of bacterial species have been sequenced, with at least 9,000 sequences completed and more than 42,000 left as "permanent" drafts (as of Sep 2016).

Most bacteria have a single circular chromosome that can range in size from only 160,000 base pairs in the endosymbiotic bacteria *Candidatus Carsonella ruddii*, to 12,200,000 base pairs in the soil-dwelling bacteria *Sorangium cellulosum*. The genes in bacterial genomes are usually a single continuous stretch of DNA and although several different types of introns do exist in bacteria, these are much rarer than in eukaryotes. Some bacteria, including the Spirochaetes of the genus *Borrelia* are a notable exception to this arrangement. *Borrelia burgdorferi*, the cause of Lyme disease, contains a single linear chromosome and several linear and circular plasmids.

Plasmids are small extra chromosomal DNAs that may contain genes for antibiotic resistance or virulence factors. Plasmids replicate independently of chromosomes, so it is possible that plasmids could be lost in bacterial cell division. Against this possibility is the fact that a single bacterium can contain hundreds of copies of a single plasmid.

GENETICS

Bacteria, as asexual organisms, inherit identical copies of their parent's genes (i.e., they are clonal). However, all bacteria can evolve by selection on changes to their genetic material DNA caused by genetic recombination or mutations. Mutations come from errors made during the replication of DNA or from exposure to mutagens. Mutation rates vary widely among different species of bacteria and even among different clones of a single species of bacteria. Genetic changes in bacterial genomes come from either random mutation during replication or stress-directed mutation, where genes involved in a particular growth-limiting process have an increased mutation rate.

DNA Transfer

Some bacteria also transfer genetic material between cells. This can occur in three main ways. First, bacteria can take up exogenous DNA from their environment, in a process called *transformation*. Genes can also be transferred by the process of *transduction*, when the integration of a bacteriophage introduces foreign DNA into the chromosome. The third method of gene transfer is *conjugation*, whereby DNA is transferred through direct cell contact.

Transduction of bacterial genes by bacteriophage appears to be a consequence of infrequent errors during intracellular assembly of virus particles, rather than a bacterial adaptation. Conjugation, in the much-studied E. coli system is determined by plasmid genes, and is an adaptation for transferring copies of the plasmid from one bacterial host to another. It is seldom that a conjugative plasmid integrates into the host bacterial chromosome, and subsequently transfers part of the host bacterial DNA to another bacterium. Plasmid-mediated transfer of host bacterial DNA also appears to be an accidental process rather than a bacterial adaptation.

Transformation, unlike transduction or conjugation, depends on numerous bacterial gene products that specifically interact to perform this complex process, and thus transformation is clearly a bacterial adaptation for DNA transfer. In order for a bacterium to bind, take up and recombine donor DNA into its own chromosome, it must first enter a special physiological state termed competence (see Natural competence). In *Bacillus subtilis*, about 40 genes are required for the development of competence. The length of DNA transferred during *B. subtilis* transformation can be between a third of a chromosome up to the whole chromosome. Transformation appears to be common among bacterial species, and thus far at least 60 species are known to have the natural ability to become competent for transformation. The development of competence in nature is usually associated with stressful environmental conditions, and seems to be an adaptation for facilitating repair of DNA damage in recipient cells.

In ordinary circumstances, transduction, conjugation, and transformation involve transfer of DNA between individual bacteria of the same species, but occasionally transfer may occur between individuals of different bacterial species and this may have significant consequences, such as the transfer of antibiotic resistance. In such cases, gene acquisition from other bacteria or the environment is called *horizontal gene transfer* and may be common under natural conditions. Gene transfer is particularly important in antibiotic resistance as it allows the rapid transfer of resistance genes between different pathogens.

Bacteriophages

Bacteriophages are viruses that infect bacteria. Many types of bacteriophage exist, some simply infect and lyse their host bacteria, while others insert into the bacterial chromosome. A bacteriophage can contain genes that contribute to its host's phenotype: for example, in the evolution of *Escherichia coli* O157:H7 and *Clostridium botulinum*, the toxin genes in an integrated phage converted a harmless ancestral bacterium into a lethal pathogen. Bacteria resist phage infection through restriction modification systems that degrade foreign DNA, and a system that uses CRISPR sequences to retain fragments of the genomes of phage that the bacteria have come into contact with in the past, which allows them to block virus replication through a form of RNA interference. This CRISPR system provides bacteria with acquired immunity to infection.

BEHAVIOUR

Secretion

Bacteria frequently secrete chemicals into their environment in order to modify it favourably. The secretions are often proteins and may act as enzymes that digest some form of food in the environment.

Bioluminescence

A few bacteria have chemical systems that generate light. This bioluminescence often occurs in bacteria that live in association with fish, and the light probably serves to attract fish or other large animals.

Multicellularity

Bacteria often function as multicellular aggregates known as biofilms, exchanging a variety of molecular signals for inter-cell communication, and engaging in coordinated multicellular behaviour.

The communal benefits of multicellular cooperation include a cellular division of labour, accessing resources that cannot effectively be used by single cells, collectively defending against antagonists, and optimising population survival by differentiating into distinct cell types. For example, bacteria in biofilms can have more than 500 times increased resistance to antibacterial agents than individual "planktonic" bacteria of the same species.

One type of inter-cellular communication by a molecular signal is called quorum sensing, which serves the purpose of determining whether there is a local population density that is sufficiently high that it is productive to invest in processes that are only successful if large numbers of similar organisms behave similarly, as in excreting digestive enzymes or emitting light.

Quorum sensing allows bacteria to coordinate gene expression, and enables them to produce, release and detect autoinducers or pheromones which accumulate with the growth in cell population.

Movement

Many bacteria can move using a variety of mechanisms: flagella are used for swimming through fluids; bacterial gliding and twitching motility move bacteria across surfaces; and changes of buoyancy allow vertical motion.

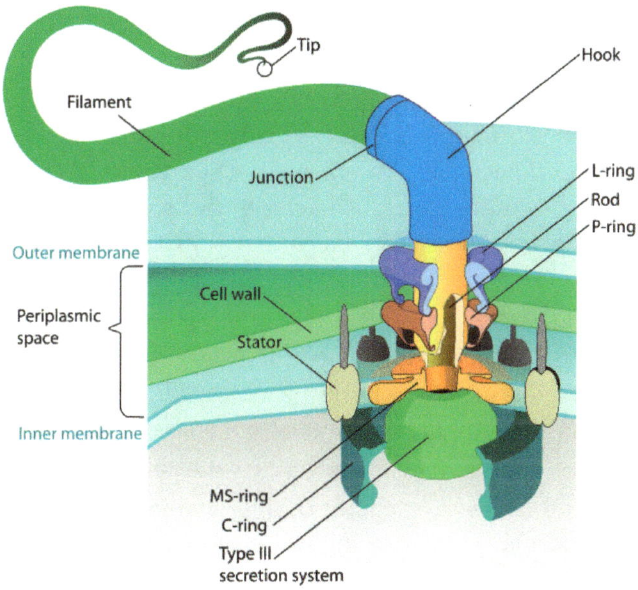

Flagellum of gram-negative bacteria. The base drives the rotation of the hook and filament.

Swimming bacteria frequently move near 10 body lengths per second and a few as fast as 100. This makes them at least as fast as fish, on a relative scale.

In bacterial gliding and twitching motility, bacteria use their *type IV pili* as a grappling hook, repeatedly extending it, anchoring it and then retracting it with remarkable force (>80 pN).

"Our observations redefine twitching motility as a rapid, highly organized mechanism of bacterial translocation by which Pseudomonas aeruginosa can disperse itself over large areas to colonize new territories. It is also now clear, both morphologically and genetically, that twitching motility and social gliding motility, such as occurs in Myxococcus xanthus, are essentially the same process.

—A re-examination of twitching motility in Pseudomonas aeruginosa – Semmler, Whitchurch & Mattick (1999)

Flagella are semi-rigid cylindrical structures that are rotated and function much like the propeller on a ship. Objects as small as bacteria operate a low Reynolds number and cylindrical forms are more efficient than the flat, paddle-like, forms appropriate at human-size scale.

Bacterial species differ in the number and arrangement of flagella on their surface; some have a single flagellum (*monotrichous*), a flagellum at each end (*amphitrichous*), clusters of flagella at the poles of the cell (*lophotrichous*), while others have flagella distributed over the entire surface of the cell (*peritrichous*). The bacterial flagella is the best-understood motility structure in any organism and is made of about 20 proteins, with approximately another 30 proteins required for its regulation and assembly. The flagellum is a rotating structure driven by a reversible motor at the base that uses the electrochemical gradient across the membrane for power. This motor drives the motion of the filament, which acts as a propeller.

Many bacteria (such as *E. coli*) have two distinct modes of movement: forward movement (swimming) and tumbling. The tumbling allows them to reorient and makes their movement a three-dimensional random walk. (See external links below for link to videos.) The flagella of a unique group of bacteria, the spirochaetes, are found between two membranes in the periplasmic space. They have a distinctive helical body that twists about as it moves.

Motile bacteria are attracted or repelled by certain stimuli in behaviours called taxes: these include chemotaxis, phototaxis, energy taxis, and magnetotaxis. In one peculiar group, the myxobacteria, individual bacteria move together to form waves of cells that then differentiate to form fruiting bodies containing spores. The myxobacteria move only when on solid surfaces, unlike *E. coli*, which is motile in liquid or solid media.

Several *Listeria* and *Shigella* species move inside host cells by usurping the cytoskeleton, which is normally used to move organelles inside the cell. By promoting actin polymerisation at one pole of their cells, they can form a kind of tail that pushes them through the host cell's cytoplasm.

CLASSIFICATION AND IDENTIFICATION

Classification seeks to describe the diversity of bacterial species by naming and grouping organisms based on similarities. Bacteria can be classified on the basis of cell structure, cellular metabolism or on differences in cell components, such as DNA, fatty acids, pigments, antigens and quinones. While these schemes allowed the identification and classification of bacterial strains, it was unclear whether these differences represented variation between distinct species or between strains of the same species. This uncertainty was due to the lack of distinctive structures in most bacteria, as well as lateral gene transfer between unrelated species. Due to lateral gene transfer, some closely related bacteria can have very different morphologies and metabolisms. To overcome this uncertainty, modern bacterial classification emphasises molecular systematics, using genetic techniques such as guanine cytosine ratio determination, genome-genome hybridisation, as well as sequencing genes that have not undergone extensive lateral gene transfer, such as the rRNA gene. Classification of bacteria is determined by publication in the International Journal of Systematic Bacteriology, and Bergey's Manual of Systematic Bacteriology. The International Committee on Systematic Bacteriology (ICSB) maintains international rules for the naming of bacteria and taxonomic categories and for the ranking of them in the International Code of Nomenclature of Bacteria.

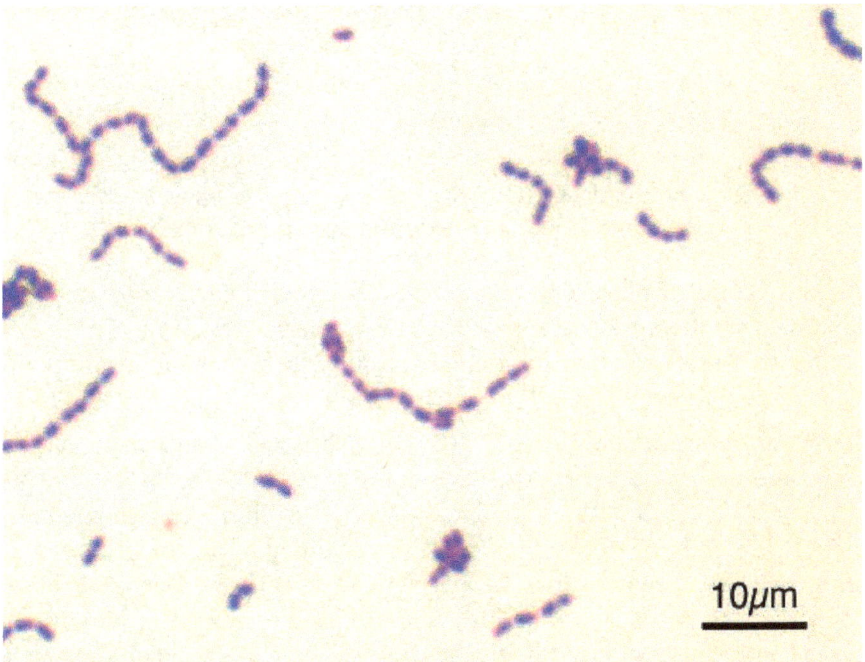

Streptococcus mutans visualised with a Gram stain

The term bacteria was traditionally applied to all microscopic, single-cell prokaryotes. However, molecular systematics showed prokaryotic life to consist of two separate domains, originally called *Eubacteria* and *Archaebacteria*, but now called *Bacteria* and *Archaea* that evolved independently from an ancient common ancestor. The archaea and eukaryotes are more closely related to each other than either is to the bacteria. These two domains, along with Eukarya, are the basis of the three-domain system, which is currently the most widely used classification system in microbiolology. However, due to the relatively recent introduction of molecular systematics and a rapid increase in the number of genome sequences that are available, bacterial classification remains a changing and expanding field. For example, a few biologists argue that the Archaea and Eukaryotes evolved from gram-positive bacteria.

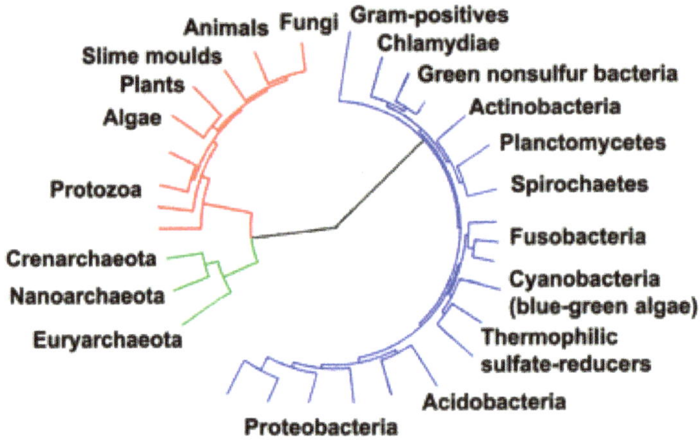

Phylogenetic tree showing the diversity of bacteria, compared to other organisms. Eukaryotes are coloured red, archaea green and bacteria blue.

The identification of bacteria in the laboratory is particularly relevant in medicine, where the correct treatment is determined by the bacterial species causing an infection. Consequently, the need to identify human pathogens was a major impetus for the development of techniques to identify bacteria.

The *Gram stain*, developed in 1884 by Hans Christian Gram, characterises bacteria based on the structural characteristics of their cell walls. The thick layers of peptidoglycan in the "gram-positive" cell wall stain purple, while the thin "gram-negative" cell wall appears pink. By combining morphology and Gram-staining, most bacteria can be classified as belonging to one of four groups (gram-positive cocci, gram-positive bacilli, gram-negative cocci and gram-negative bacilli). Some organisms are best identified by stains other than the Gram stain, particularly mycobacteria or *Nocardia*, which show acid-fastness on Ziehl–Neelsen or similar stains. Other organisms may need to be identified by their growth in special media, or by other techniques, such as serology.

Culture techniques are designed to promote the growth and identify particular bacteria, while restricting the growth of the other bacteria in the sample. Often these techniques are designed for specific specimens; for example, a sputum sample will be treated to identify organisms that cause pneumonia, while stool specimens are cultured on selective media to identify organisms that cause diarrhoea, while preventing growth of non-pathogenic bacteria. Specimens that are normally sterile, such as blood, urine or spinal fluid, are cultured under conditions designed to grow all possible organisms. Once a pathogenic organism has been isolated, it can be further characterised by its morphology, growth patterns (such as aerobic or anaerobic growth), patterns of hemolysis, and staining.

As with bacterial classification, identification of bacteria is increasingly using molecular methods. Diagnostics using DNA-based tools, such as polymerase chain reaction, are increasingly popular due to their specificity and speed, compared to culture-based methods. These methods also allow the detection and identification of viable but nonculturable cells that are metabolically active but non-dividing. However, even using these improved methods, the total number of bacterial species is not known and cannot even be estimated with any certainty. Following present classification, there are a little less than 9,300 known species of prokaryotes, which includes bacteria and archaea; but attempts to estimate the true number of bacterial diversity have ranged from 10^7 to 10^9 total species – and even these diverse estimates may be off by many orders of magnitude.

INTERACTIONS WITH OTHER ORGANISMS

Despite their apparent simplicity, bacteria can form complex associations with other organisms. These symbiotic associations can be divided into parasitism, mutualism and commensalism. Due to their small size, commensal bacteria are ubiquitous and grow on animals and plants exactly as they will grow on any other surface. However, their growth can be increased by warmth and sweat, and large populations of these organisms in humans are the cause of body odour.

Predators

Some species of bacteria kill and then consume other microorganisms, these species are called *predatory bacteria*. These include organisms such as *Myxococcus xanthus*, which forms swarms of cells that kill and digest any bacteria they encounter. Other bacterial predators either attach to their prey in order to digest them and absorb nutrients, such as *Vampirovibrio chlorellavorus*, or invade another cell and multiply inside the cytosol, such as *Daptobacter*. These predatory bacteria are thought to have evolved from saprophages that consumed dead microorganisms, through adaptations that allowed them to entrap and kill other organisms.

Mutualists

Certain bacteria form close spatial associations that are essential for their survival. One such mutualistic association, called interspecies hydrogen transfer, occurs between clusters of anaerobic bacteria that consume organic acids, such as butyric acid or propionic acid, and produce hydrogen, and methanogenic Archaea that consume hydrogen. The bacteria in this association are unable to consume the organic acids as this reaction produces hydrogen that accumulates in their surroundings. Only the intimate association with the hydrogen-consuming Archaea keeps the hydrogen concentration low enough to allow the bacteria to grow.

In soil, microorganisms that reside in the rhizosphere (a zone that includes the root surface and the soil that adheres to the root after gentle shaking) carry out nitrogen fixation, converting nitrogen gas to nitrogenous compounds. This serves to provide an easily absorbable form of nitrogen for many plants, which cannot fix nitrogen themselves.

Many other bacteria are found as symbionts in humans and other organisms. For example, the presence of over 1,000 bacterial species in the normal human gut flora of the intestines can contribute to gut immunity, synthesise vitamins, such as folic acid, vitamin K and biotin, convert sugars to lactic acid (see *Lactobacillus*), as well as fermenting complex undigestible carbohydrates. The presence of this gut flora also inhibits the growth of potentially pathogenic bacteria (usually through competitive exclusion) and these beneficial bacteria are consequently sold as probiotic dietary supplements.

Pathogens

If bacteria form a parasitic association with other organisms, they are classed as pathogens. Pathogenic bacteria are a major cause of human death and disease and cause infections such as tetanus, typhoid fever, diphtheria, syphilis, cholera, foodborne illness, leprosy and tuberculosis. A pathogenic cause for a known medical disease may only be discovered many years after, as was the case with *Helicobacter pylori* and peptic ulcer disease. Bacterial diseases are also important in agriculture, with bacteria causing leaf spot, fire blight and wilts in plants, as well as Johne's disease, mastitis, salmonella and anthrax in farm animals.

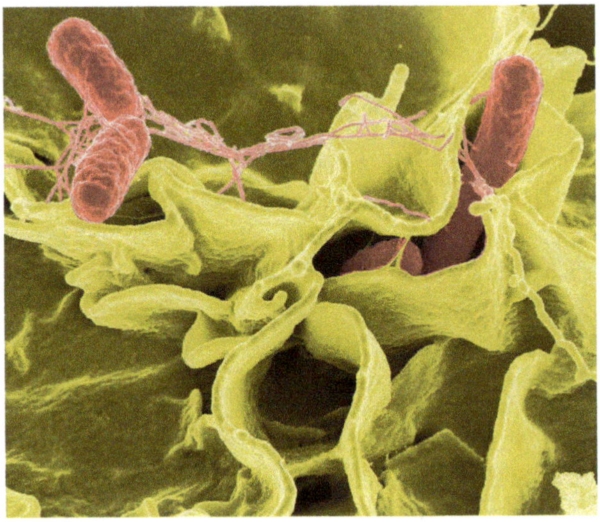

Colour-enhanced scanning electron micrograph showing *Salmonella typhimurium* (red) invading cultured human cells

Each species of pathogen has a characteristic spectrum of interactions with its human hosts. Some organisms, such as *Staphylococcus* or *Streptococcus*, can cause skin infections, pneumonia, meningitis and even overwhelming sepsis, a systemic inflammatory response producing shock, massive vasodilation and death. Yet these organisms are also part of the normal human flora and usually exist on the skin or in the nose without causing any disease at all. Other organisms invariably cause disease in humans, such as the Rickettsia, which are obligate intracellular parasites able to grow and reproduce only within the cells of other organisms. One species of Rickettsia causes typhus, while another causes Rocky Mountain spotted fever. *Chlamydia*, another phylum of obligate intracellular parasites, contains species that can cause pneumonia, or urinary tract infection and may be involved in coronary heart disease. Finally, some species, such as *Pseudomonas aeruginosa*, *Burkholderia cenocepacia*, and *Mycobacterium avium*, are opportunistic pathogens and cause disease mainly in people suffering from immunosuppression or cystic fibrosis.

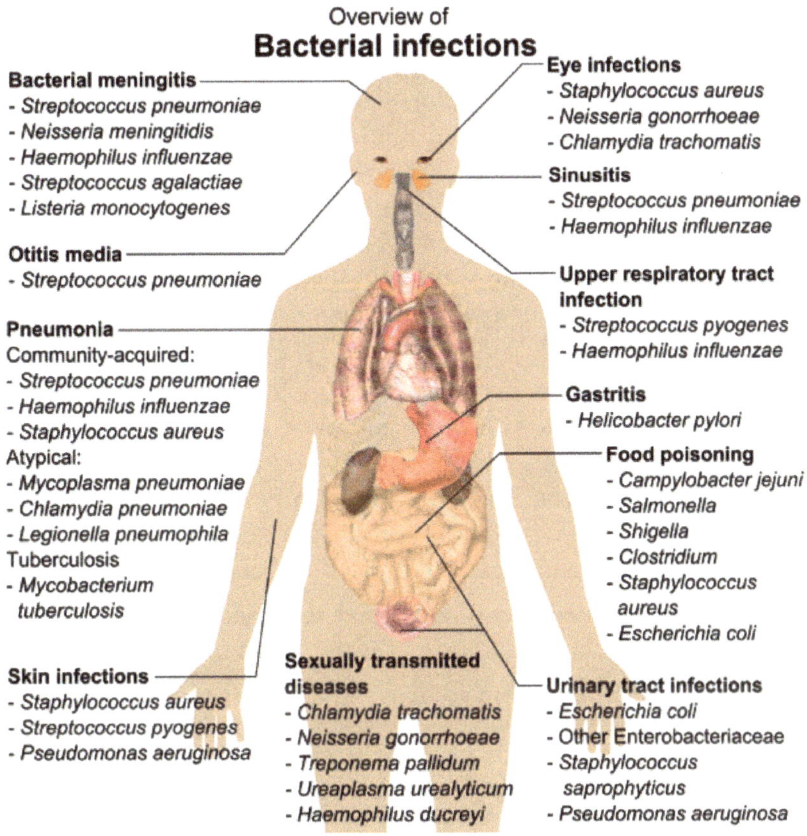

Overview of bacterial infections and main species involved.

Bacterial infections may be treated with antibiotics, which are classified as bacteriocidal if they kill bacteria, or bacteriostatic if they just prevent bacterial growth. There are many types of antibiotics and each class inhibits a process that is different in the pathogen from that found in the host. An example of how antibiotics produce selective toxicity are chloramphenicol and puromycin, which inhibit the bacterial ribosome, but not the structurally different eukaryotic ribosome. Antibiotics are used both in treating human disease and in intensive farming to promote animal growth, where they may be contributing to the rapid development of antibiotic resistance in bacterial populations.Infections can be prevented by antiseptic measures such as sterilising the skin prior to piercing it with the needle of a syringe, and by proper care of indwelling catheters. Surgical and dental instruments are also sterilised to prevent contamination by bacteria. Disinfectants such as bleach are used to kill bacteria or other pathogens on surfaces to prevent contamination and further reduce the risk of infection.

Significance in Technology and Industry

Bacteria, often lactic acid bacteria, such as *Lactobacillus* and *Lactococcus*, in combination with yeasts and moulds, have been used for thousands of years in the preparation of fermented foods, such as cheese, pickles, soy sauce, sauerkraut, vinegar, wine and yogurt.

The ability of bacteria to degrade a variety of organic compounds is remarkable and has been used in waste processing and bioremediation. Bacteria capable of digesting the hydrocarbons in petroleum are often used to clean up oil spills. Fertiliser was added to some of the beaches in Prince William Sound in an attempt to promote the growth of these naturally occurring bacteria after the 1989 *Exxon Valdez* oil spill. These efforts were effective on beaches that were not too thickly covered in oil. Bacteria are also used for the bioremediation of industrial toxic wastes. In the chemical industry, bacteria are most important in the production of enantiomerically pure chemicals for use as pharmaceuticals or agrichemicals.

Bacteria can also be used in the place of pesticides in the biological pest control. This commonly involves *Bacillus thuringiensis* (also called BT), a gram-positive, soil dwelling bacterium. Subspecies of this bacteria are used as a Lepidopteran-specific insecticides under trade names such as Dipel and Thuricide. Because of their specificity, these pesticides are regarded as environmentally friendly, with little or no effect on humans, wildlife, pollinators and most other beneficial insects.

Because of their ability to quickly grow and the relative ease with which they can be manipulated, bacteria are the workhorses for the fields of molecular biology, genetics and biochemistry. By making mutations in bacterial DNA and examining the resulting phenotypes, scientists can determine the function of genes, enzymes and metabolic pathways in bacteria, then apply this knowledge to more complex organisms. This aim of understanding the biochemistry of a cell reaches its most complex expression in the synthesis of huge amounts of enzyme kinetic and gene expression data into mathematical models of entire organisms. This is achievable in some well-studied bacteria, with models of *Escherichia coli* metabolism now being produced and tested. This understanding of bacterial metabolism and genetics allows the use of biotechnology to bioengineer bacteria for the production of therapeutic proteins, such as insulin, growth factors, or antibodies.

Because of their importance for research in general, samples of bacterial strains are isolated and preserved in Biological Resource Centers. This ensures the availability of the strain to scientists worldwide.

History of Bacteriology

Bacteria were first observed by the Dutch microscopist Antonie van Leeuwenhoek in 1676, using a single-lens microscope of his own design. He then published his observations in a series of letters to the Royal Society of London. Bacteria were Leeuwenhoek's most remarkable microscopic discovery. They were just at the limit of what his simple lenses could make out and, in one of the most striking hiatuses in the history of science, no one else would see them again for over a century. Only then were his by-then-largely-forgotten observations of bacteria — as opposed to his famous animalcules (spermatozoa) — taken seriously.

Antonie van Leeuwenhoek, the first microbiologist and the first person to observe bacteria using a microscope.

Christian Gottfried Ehrenberg introduced the word "bacterium" in 1828. In fact, his *Bacterium* was a genus that contained non-spore-forming rod-shaped bacteria, as opposed to *Bacillus*, a genus of spore-forming rod-shaped bacteria defined by Ehrenberg in 1835.

Louis Pasteur demonstrated in 1859 that the growth of microorganisms causes the fermentation process, and that this growth is not due to spontaneous generation. (Yeasts and moulds, commonly associated with fermentation, are not bacteria, but rather fungi.) Along with his contemporary Robert Koch, Pasteur was an early advocate of the germ theory of disease.

Robert Koch, a pioneer in medical microbiology, worked on cholera, anthrax and tuberculosis. In his research into tuberculosis Koch finally proved the germ theory, for which he received a Nobel Prize in 1905. In *Koch's postulates*, he set out criteria to test if an organism is the cause of a disease, and these postulates are still used today.

Though it was known in the nineteenth century that bacteria are the cause of many diseases, no effective antibacterial treatments were available. In 1910, Paul Ehrlich developed the first antibiotic, by changing dyes that selectively stained *Treponema pallidum* — the spirochaete that causes syphilis — into compounds that selectively killed the pathogen. Ehrlich had been awarded a 1908 Nobel Prize for his work on immunology, and pioneered the use of stains to detect and identify bacteria, with his work being the basis of the Gram stain and the Ziehl–Neelsen stain.

A major step forward in the study of bacteria came in 1977 when Carl Woese recognised that archaea have a separate line of evolutionary descent from bacteria. This new phylogenetic taxonomy depended on the sequencing of 16S ribosomal RNA, and divided prokaryotes into two evolutionary domains, as part of the three-domain system.

REFERENCES

1. Woese CR, Kandler O, Wheelis ML (1990). "Towards a natural system of organisms: proposal for the domains Archaea, Bacteria, and Eucarya". Proceedings of the National Academy of Sciences of the United States of America. 87 (12): 4576–9. Bibcode:1990PNAS...87.4576W. doi:10.1073/pnas.87.12.4576. PMC 54159Freely accessible. PMID 2112744.
2. "Bacteria (eubacteria)". Taxonomy Browser, US National Institute of Health. Retrieved 10 September 2008.
3. Woese CR, Fox GE (1977). "Phylogenetic structure of the prokaryotic domain: the primary kingdoms". Proceedings of the National Academy of Sciences of the United States of America. 74 (11): 5088–90. Bibcode:1977PNAS...74.5088W. doi:10.1073/pnas.74.11.5088. PMC 432104Freely accessible. PMID 270744.
4. Fredrickson JK, Zachara JM, Balkwill DL, Kennedy D, Li SM, Kostandarithes HM, Daly MJ, Romine MF, Brockman FJ (2004). "Geomicrobiology of high-level nuclear waste-contaminated vadose sediments at the Hanford site, Washington state". Applied and Environmental Microbiology. 70 (7): 4230–41. doi:10.1128/AEM.70.7.4230-4241.2004. PMC 444790Freely accessible. PMID 15240306.
5. Whitman WB, Coleman DC, Wiebe WJ (1998). "Prokaryotes: the unseen majority". Proceedings of the National Academy of Sciences of the United States of America. 95 (12): 6578–83. Bibcode:1998PNAS...95.6578W. doi:10.1073/pnas.95.12.6578. PMC 33863Freely accessible. PMID 9618454.
6. C.Michael Hogan. 2010. Bacteria. Encyclopedia of Earth. eds. Sidney Draggan and C.J.Cleveland, National Council for Science and the Environment, Washington DC
7. Choi CQ (17 March 2013). "Microbes Thrive in Deepest Spot on Earth". LiveScience. Retrieved 17 March 2013.
8. Glud R, Wenzhöfer F, Middelboe M, Oguri K, Turnewitsch R, Canfield DE, Kitazato H (2013). "High rates of microbial carbon turnover in sediments in the deepest oceanic trench on Earth". Nature Geoscience. 6 (4): 284–288. Bibcode:2013NatGe...6..284G. doi:10.1038/ngeo1773.
9. Oskin B (14 March 2013). "Intraterrestrials: Life Thrives in Ocean Floor". LiveScience. Retrieved 17 March 2013.
10. Rappé MS, Giovannoni SJ (2003). "The uncultured microbial majority". Annual Review of Microbiology. 57: 369–94. doi:10.1146/annurev.micro.57.030502.090759. PMID 14527284.
11. Sears CL (2005). "A dynamic partnership: celebrating our gut flora". Anaerobe. 11 (5): 247–51. doi:10.1016/j.anaerobe.2005.05.001. PMID 16701579.
12. "2002 WHO mortality data". Retrieved 20 January 2007.

13. "Metal-Mining Bacteria Are Green Chemists". Science Daily. 2 September 2010.
14. Ishige T, Honda K, Shimizu S (2005). "Whole organism biocatalysis". Current Opinion in Chemical Biology. 9 (2): 174–80. doi:10.1016/j.cbpa.2005.02.001. PMID 15811802.
15. βακτήριον. Liddell, Henry George; Scott, Robert; A Greek–English Lexicon at the Perseus Project.
16. βακτηρία in Liddell and Scott.
17. bacterium, on Oxford Dictionaries.
18. Harper, Douglas. bacteria. Online Etymology Dictionary.
19. Schopf JW (1994). Disparate rates, differing fates: tempo and mode of evolution changed from the Precambrian to the Phanerozoic. Proceedings of the National Academy of Sciences of the United States of America. 91 (15): 6735–42. Bibcode:1994PNAS...91.6735S. doi:10.1073/pnas.91.15.6735. PMC 44277Freely accessible. PMID 8041691.
20. DeLong EF, Pace NR (2001). Environmental diversity of bacteria and archaea. Syst Biol. 50 (4): 470–8. doi:10.1080/106351501750435040. PMID 12116647.
21. Brown JR, Doolittle WF (1997). Archaea and the prokaryote-to-eukaryote transition. Microbiology and Molecular Biology Reviews. 61 (4): 456–502. PMC 232621Freely accessible. PMID 9409149.
22. Poole AM, Penny D (2007). Evaluating hypotheses for the origin of eukaryotes. BioEssays. 29 (1): 74–84. doi:10.1002/bies.20516. PMID 17187354.
23. Dyall SD, Brown MT, Johnson PJ (2004). Ancient invasions: from endosymbionts to organelles. Science. 304 (5668): 253–7. Bibcode:2004Sci...304..253D. doi:10.1126/science.1094884. PMID 15073369.
24. Lang BF, Gray MW, Burger G (1999). Mitochondrial genome evolution and the origin of eukaryotes. Annu Rev Genet. 33: 351–97. doi:10.1146/annurev.genet.33.1.351. PMID 10690412.
25. McFadden GI (1999). Endosymbiosis and evolution of the plant cell. Current Opinion in Plant Biology. 2 (6): 513–9. doi:10.1016/S1369-5266(99)00025-4. PMID 10607659.
26. Schulz HN, Jorgensen BB (2001). Big bacteria. Annu Rev Microbiol. 55: 105–37. doi:10.1146/annurev.micro.55.1.105. PMID 11544351.
27. Williams C (2011). Who are you calling simple?. New Scientist. 211 (2821): 38–41. doi:10.1016/S0262-4079(11)61709-0.
28. Robertson J, Gomersall M, Gill P (1975). Mycoplasma hominis: growth, reproduction, and isolation of small viable cells. J Bacteriol. 124 (2): 1007–18. PMC 235991Freely accessible. PMID 1102522.
29. Velimirov B (2001). Nanobacteria, Ultramicrobacteria and Starvation Forms: A Search for the Smallest Metabolizing Bacterium. Microbes and Environments. 16 (2): 67–77. doi:10.1264/jsme2.2001.67.

30. Dusenbery, David B. (2009). Living at Micro Scale, pp. 20–25. Harvard University Press, Cambridge, Mass. ISBN 978-0-674-03116-6.

31. Fritz I, Strömpl C, Abraham WR (2004). Phylogenetic relationships of the genera Stella, Labrys and Angulomicrobium within the 'Alphaproteobacteria' and description of Angulomicrobium amanitiforme sp. nov. Int J Syst Evol Microbiol. 54 (Pt 3): 651–7. doi:10.1099/ijs.0.02746-0. PMID 15143003.

32. Wanger G, Onstott TC, Southam G (2008). Stars of the terrestrial deep subsurface: A novel 'star-shaped' bacterial morphotype from a South African platinum mine. Geobiology. 6 (3): 325–30. doi:10.1111/j.1472-4669.2008.00163.x. PMID 18498531.

33. Cabeen MT, Jacobs-Wagner C (2005). Bacterial cell shape. Nature Reviews Microbiology. 3 (8): 601–10. doi:10.1038/nrmicro1205. PMID 16012516.

34. Young KD (2006). The selective value of bacterial shape. Microbiol Mol Biol Rev. 70 (3): 660–703. doi:10.1128/MMBR.00001-06. PMC 1594593Freely accessible. PMID 16959965.

35. Douwes KE, Schmalzbauer E, Linde HJ, Reisberger EM, Fleischer K, Lehn N, Landthaler M, Vogt T (2003). Branched filaments no fungus, ovoid bodies no bacteria: Two unusual cases of mycetoma. J Am Acad Dermatol. 49 (2 Suppl Case Reports): S170–3. doi:10.1067/mjd.2003.302. PMID 12894113.

36. Donlan RM (2002). Biofilms: microbial life on surfaces. Emerg Infect Dis. 8 (9): 881–90. doi:10.3201/eid0809.020063. PMC 2732559Freely accessible. PMID 12194761.

37. Branda SS, Vik S, Friedman L, Kolter R (2005). Biofilms: the matrix revisited. Trends Microbiol. 13 (1): 20–6. doi:10.1016/j.tim.2004.11.006. PMID 15639628.

38. Davey ME, O'toole GA (2000). Microbial biofilms: from ecology to molecular genetics. Microbiol Mol Biol Rev. 64 (4): 847–67. doi:10.1128/MMBR.64.4.847-867.2000. PMC 99016Freely accessible. PMID 11104821.

39. Donlan RM, Costerton JW (2002). Biofilms: survival mechanisms of clinically relevant microorganisms. Clin Microbiol Rev. 15 (2): 167–93. doi:10.1128/CMR.15.2.167-193.2002. PMC 118068Freely accessible. PMID 11932229.

40. Shimkets LJ (1999). Intercellular signaling during fruiting-body development of Myxococcus xanthus. Annu Rev Microbiol. 53: 525–49. doi:10.1146/annurev.micro.53.1.525. PMID 10547700.

41. Kaiser D (2004). Signaling in myxobacteria. Annu Rev Microbiol. 58: 75–98. doi:10.1146/annurev.micro.58.030603.123620. PMID 15487930.

42. Berg JM, Tymoczko JL, Stryer L (2002). Molecular Cell Biology (5th ed.). WH Freeman. ISBN 0-7167-4955-6.

43. Gitai Z (2005). The new bacterial cell biology: moving parts and subcellular architecture. Cell. 120 (5): 577–86. doi:10.1016/j.cell.2005.02.026. PMID 15766522.

44. Shih YL, Rothfield L (2006). The bacterial cytoskeleton. Microbiology and Molecular Biology Reviews. 70 (3): 729–54. doi:10.1128/MMBR.00017-06.

PMC 1594594Freely accessible. PMID 16959967.

45. Norris V, den Blaauwen T, Cabin-Flaman A, Doi RH, Harshey R, Janniere L, Jimenez-Sanchez A, Jin DJ, Levin PA, Mileykovskaya E, Minsky A, Saier M, Skarstad K (2007). Functional taxonomy of bacterial hyperstructures. Microbiology and Molecular Biology Reviews. 71 (1): 230–53. doi:10.1128/MMBR.00035-06. PMC 1847379Freely accessible. PMID 17347523.

46. Kerfeld CA, Sawaya MR, Tanaka S, Nguyen CV, Phillips M, Beeby M, Yeates TO (2005). Protein structures forming the shell of primitive bacterial organelles. Science. 309 (5736): 936–8. Bibcode:2005Sci...309..936K. doi:10.1126/science.1113397. PMID 16081736.

47. Bobik TA (2007). Bacterial Microcompartments (PDF). Microbe. Am Soc Microbiol. 2: 25–31. doi:10.1128/microbe.2.25.1. Archived from the original (PDF) on 2 August 2008.

48. Bobik TA (2006). Polyhedral organelles compartmenting bacterial metabolic processes. Applied Microbiology and Biotechnology. 70 (5): 517–25. doi:10.1007/s00253-005-0295-0. PMID 16525780.

49. Yeates TO, Kerfeld CA, Heinhorst S, Cannon GC, Shively JM (2008). Protein-based organelles in bacteria: carboxysomes and related microcompartments. Nature Reviews Microbiology. 6 (9): 681–91. doi:10.1038/nrmicro1913. PMID 18679172.

50. Harold FM (1972). Conservation and transformation of energy by bacterial membranes. Bacteriological Reviews. 36 (2): 172–230. PMC 408323Freely accessible. PMID 4261111.

51. Bryant DA, Frigaard NU (2006). Prokaryotic photosynthesis and phototrophy illuminated. Trends Microbiol. 14 (11): 488–96. doi:10.1016/j.tim.2006.09.001. PMID 16997562.

52. Psencík J, Ikonen TP, Laurinmäki P, Merckel MC, Butcher SJ, Serimaa RE, Tuma R (2004). Lamellar organization of pigments in chlorosomes, the light harvesting complexes of green photosynthetic bacteria. Biophys. J. 87 (2): 1165–72. Bibcode:2004BpJ....87.1165P. doi:10.1529/biophysj.104.040956. PMC 1304455Freely accessible. PMID 15298919.

53. Tanaka S, Kerfeld CA, Sawaya MR, Cai F, Heinhorst S, Cannon GC, Yeates TO (2008). Atomic-level models of the bacterial carboxysome shell. Science. 319 (5866): 1083–6. Bibcode:2008Sci...319.1083T. doi:10.1126/science.1151458. PMID 18292340.

54. Thanbichler M, Wang SC, Shapiro L (2005). The bacterial nucleoid: a highly organized and dynamic structure. J Cell Biochem. 96 (3): 506–21. doi:10.1002/jcb.20519. PMID 15988757.

55. Fuerst JA (2005). Intracellular compartmentation in planctomycetes. Annu Rev Microbiol. 59: 299–328. doi:10.1146/annurev.micro.59.030804.121258. PMID 15910279.

56. Fieseler L, Horn M, Wagner M, Hentschel U (June 2004). Discovery of the novel candidate phylum Poribacteria in marine sponges. Applied and Environmental

Microbiology. 70 (6): 3724–32. doi:10.1128/AEM.70.6.3724-3732.2004. PMC 427773Freely accessible. PMID 15184179.

57. Poehlsgaard J, Douthwaite S (2005). The bacterial ribosome as a target for antibiotics. Nature Reviews Microbiology. 3 (11): 870–81. doi:10.1038/nrmicro1265. PMID 16261170.

58. Yeo M, Chater K (2005). The interplay of glycogen metabolism and differentiation provides an insight into the developmental biology of Streptomyces coelicolor. Microbiology. 151 (Pt 3): 855–61. doi:10.1099/mic.0.27428-0. PMID 15758231.

59. Shiba T, Tsutsumi K, Ishige K, Noguchi T (2000). Inorganic polyphosphate and polyphosphate kinase: their novel biological functions and applications. Biochemistry (Mosc). 65 (3): 315–23. PMID 10739474.

60. Brune DC (1995). Isolation and characterization of sulfur globule proteins from Chromatium vinosum and Thiocapsa roseopersicina. Archives of Microbiology. 163 (6): 391–9. doi:10.1007/BF00272127. PMID 7575095.

61. Kadouri D, Jurkevitch E, Okon Y, Castro-Sowinski S (2005). Ecological and agricultural significance of bacterial polyhydroxyalkanoates. Critical Reviews in Microbiology. 31 (2): 55–67. doi:10.1080/10408410590899228. PMID 15986831.

62. Walsby AE (1994). Gas vesicles. Microbiological Reviews. 58 (1): 94–144. PMC 372955Freely accessible. PMID 8177173.

63. van Heijenoort J (2001). Formation of the glycan chains in the synthesis of bacterial peptidoglycan. Glycobiology. 11 (3): 25R–36R. doi:10.1093/glycob/11.3.25R. PMID 11320055.

64. Koch AL (2003). Bacterial wall as target for attack: past, present, and future research. Clin Microbiol Rev. 16 (4): 673–87. doi:10.1128/CMR.16.4.673-687.2003. PMC 207114Freely accessible. PMID 14557293.

65. Gram, HC (1884). Über die isolierte Färbung der Schizomyceten in Schnitt- und Trockenpräparaten. Fortschr. Med. 2: 185–189.

66. Hugenholtz P (2002). Exploring prokaryotic diversity in the genomic era. Genome Biology. 3 (2): reviews0003.1–reviews0003.8. doi:10.1186/gb-2002-3-2-reviews0003. PMC 139013Freely accessible. PMID 11864374.

67. Walsh FM, Amyes SG (2004). Microbiology and drug resistance mechanisms of fully resistant pathogens. Current Opinion in Microbiology. 7 (5): 439–44. doi:10.1016/j.mib.2004.08.007. PMID 15451497.

68. Engelhardt H, Peters J (1998). Structural research on surface layers: a focus on stability, surface layer homology domains, and surface layer-cell wall interactions. J Struct Biol. 124 (2–3): 276–302. doi:10.1006/jsbi.1998.4070. PMID 10049812.

69. Beveridge TJ, Pouwels PH, Sára M, Kotiranta A, Lounatmaa K, Kari K, Kerosuo E, Haapasalo M, Egelseer EM, Schocher I, Sleytr UB, Morelli L, Callegari ML, Nomellini JF, Bingle WH, Smit J, Leibovitz E, Lemaire M, Miras I, Salamitou S, Béguin P, Ohayon H, Gounon P, Matuschek M, Koval SF (1997). Functions of S-layers. FEMS Microbiol Rev. 20 (1–2): 99–149. doi:10.1016/S0168-6445(97)00043-0. PMID 9276929.

70. Kojima S, Blair DF (2004). The bacterial flagellar motor: structure and function of a complex molecular machine. Int Rev Cytol. International Review of Cytology. 233: 93–134. doi:10.1016/S0074-7696(04)33003-2. ISBN 978-0-12-364637-8. PMID 15037363.

71. Beachey EH (1981). Bacterial adherence: adhesin-receptor interactions mediating the attachment of bacteria to mucosal surface. J Infect Dis. 143 (3): 325–45. doi:10.1093/infdis/143.3.325. PMID 7014727.

72. Silverman PM (1997). Towards a structural biology of bacterial conjugation. Mol Microbiol. 23 (3): 423–9. doi:10.1046/j.1365-2958.1997.2411604.x. PMID 9044277.

73. Stokes RW, Norris-Jones R, Brooks DE, Beveridge TJ, Doxsee D, Thorson LM (2004). The glycan-rich outer layer of the cell wall of Mycobacterium tuberculosis acts as an antiphagocytic capsule limiting the association of the bacterium with macrophages. Infect Immun. 72 (10): 5676–86. doi:10.1128/IAI.72.10.5676-5686.2004. PMC 517526Freely accessible. PMID 15385466.

74. Daffé M, Etienne G (1999). The capsule of Mycobacterium tuberculosis and its implications for pathogenicity. Tuber Lung Dis. 79 (3): 153–69. doi:10.1054/tuld.1998.0200. PMID 10656114.

75. Finlay BB, Falkow S (1997). Common themes in microbial pathogenicity revisited. Microbiology and Molecular Biology Reviews. 61 (2): 136–69. PMC 232605Freely accessible. PMID 9184008.

76. Nicholson WL, Munakata N, Horneck G, Melosh HJ, Setlow P (2000). Resistance of Bacillus endospores to extreme terrestrial and extraterrestrial environments. Microbiology and Molecular Biology Reviews. 64 (3): 548–72. doi:10.1128/MMBR.64.3.548-572.2000. PMC 99004Freely accessible. PMID 10974126.

77. Siunov AV, Nikitin DV, Suzina NE, Dmitriev VV, Kuzmin NP, Duda VI (1999). Phylogenetic status of Anaerobacter polyendosporus, an anaerobic, polysporogenic bacterium (PDF). Int J Syst Bacteriol. 49 (3): 1119–24. doi:10.1099/00207713-49-3-1119. PMID 10425769.

78. Nicholson WL, Fajardo-Cavazos P, Rebeil R, Slieman TA, Riesenman PJ, Law JF, Xue Y (2002). Bacterial endospores and their significance in stress resistance. Antonie Van Leeuwenhoek. 81 (1–4): 27–32. doi:10.1023/A:1020561122764. PMID 12448702.

79. Vreeland RH, Rosenzweig WD, Powers DW (2000). Isolation of a 250 million-year-old halotolerant bacterium from a primary salt crystal. Nature. 407 (6806): 897–900. Bibcode:2000Natur.407..897V. doi:10.1038/35038060. PMID 11057666.

80. Cano RJ, Borucki MK (1995). Revival and identification of bacterial spores in 25- to 40-million-year-old Dominican amber. Science. 268 (5213): 1060–4. Bibcode:1995Sci...268.1060C. doi:10.1126/science.7538699. PMID 7538699.

81. Nicholson WL, Schuerger AC, Setlow P (2005). The solar UV environment and bacterial spore UV resistance: considerations for Earth-to-Mars transport by natural processes and human spaceflight. Mutat Res. 571 (1–2): 249–64. doi:10.1016/j.mrfmmm.2004.10.012. PMID 15748651.

82. BBC Staff (23 August 2011). Impacts 'more likely' to have spread life from Earth. BBC. Retrieved 24 August 2011.
83. Hatheway CL (1990). Toxigenic clostridia. Clinical Microbiology Reviews. 3 (1): 66–98. PMC 358141Freely accessible. PMID 2404569.
84. Nealson KH (1999). Post-Viking microbiology: new approaches, new data, new insights. Origins of Life and Evolution of Biospheres. 29 (1): 73–93. doi:10.1023/A:1006515817767. PMID 11536899.
85. Xu J (2006). Microbial ecology in the age of genomics and metagenomics: concepts, tools, and recent advances. Mol Ecol. 15 (7): 1713–31. doi:10.1111/j.1365-294X.2006.02882.x. PMID 16689892.
86. Zillig W (1991). Comparative biochemistry of Archaea and Bacteria. Current Opinion in Genetics & Development. 1 (4): 544–51. doi:10.1016/S0959-437X(05)80206-0. PMID 1822288.
87. Hellingwerf KJ, Crielaard W, Hoff WD, Matthijs HC, Mur LR, van Rotterdam BJ (1994). Photobiology of bacteria. Antonie Van Leeuwenhoek. 65 (4): 331–47. doi:10.1007/BF00872217. PMID 7832590.
88. Zumft WG (1 December 1997). Cell biology and molecular basis of denitrification. Microbiol Mol Biol Rev. 61 (4): 533–616. PMC 232623Freely accessible. PMID 9409151.
89. Drake HL, Daniel SL, Küsel K, Matthies C, Kuhner C, Braus-Stromeyer S (1997). Acetogenic bacteria: what are the in situ consequences of their diverse metabolic versatilities?. BioFactors. 6 (1): 13–24. doi:10.1002/biof.5520060103. PMID 9233536.
90. Morel FM, Kraepiel AM, Amyot M (1998). The chemical cycle and bioaccumulation of mercury. Annual Review of Ecology and Systematics. 29: 543–566. doi:10.1146/annurev.ecolsys.29.1.543.
91. Dalton H (2005). The Leeuwenhoek Lecture 2000 the natural and unnatural history of methane-oxidizing bacteria. Philosophical Transactions of the Royal Society B. 360 (1458): 1207–22. doi:10.1098/rstb.2005.1657. PMC 1569495Freely accessible. PMID 16147517.
92. Zehr JP, Jenkins BD, Short SM, Steward GF (2003). Nitrogenase gene diversity and microbial community structure: a cross-system comparison. Environ Microbiol. 5 (7): 539–54. doi:10.1046/j.1462-2920.2003.00451.x. PMID 12823187.
93. Lonhienne TG, Sagulenko E, Webb RI, Lee KC, Franke J, Devos DP, Nouwens A, Carroll BJ, Fuerst JA (2010). Endocytosis-like protein uptake in the bacterium Gemmata obscuriglobus. Proceedings of the National Academy of Sciences of the United States of America. 107 (29): 12883–12888. Bibcode:2010PNAS..10712883L. doi:10.1073/pnas.1001085107. PMC 2919973Freely accessible. PMID 20566852.
94. Koch AL (2002). Control of the bacterial cell cycle by cytoplasmic growth. Crit Rev Microbiol. 28 (1): 61–77. doi:10.1080/1040-840291046696. PMID 12003041.
95. Eagon RG (1962). Pseudomonas natriegens, a marine bacterium with a generation time of less than 10 minutes. Journal of Bacteriology. 83 (4): 736–7. PMC

279347Freely accessible. PMID 13888946.

96. Stewart EJ, Madden R, Paul G, Taddei F (2005). Aging and death in an organism that reproduces by morphologically symmetric division. PLoS Biol. 3 (2): e45. doi:10.1371/journal.pbio.0030045. PMC 546039Freely accessible. PMID 15685293.

97. Thomson RB, Bertram H (2001). Laboratory diagnosis of central nervous system infections. Infectious Disease Clinics of North America. 15 (4): 1047–71. doi:10.1016/S0891-5520(05)70186-0. PMID 11780267.

98. Paerl HW, Fulton RS, Moisander PH, Dyble J (2001). Harmful freshwater algal blooms, with an emphasis on cyanobacteria. ScientificWorldJournal. 1: 76–113. doi:10.1100/tsw.2001.16. PMID 12805693.

99. Challis GL, Hopwood DA (2003). Synergy and contingency as driving forces for the evolution of multiple secondary metabolite production by Streptomyces species. Proceedings of the National Academy of Sciences of the United States of America. 100 Suppl 2 (90002): 14555–61. Bibcode:2003PNAS..10014555C. doi:10.1073/pnas.1934677100. PMC 304118Freely accessible. PMID 12970466.

100. Kooijman SA, Auger P, Poggiale JC, Kooi BW (2003). Quantitative steps in symbiogenesis and the evolution of homeostasis. Biol Rev Camb Philos Soc. 78 (3): 435–63. doi:10.1017/S1464793102006127. PMID 14558592.

101. Prats C, López D, Giró A, Ferrer J, Valls J (2006). Individual-based modelling of bacterial cultures to study the microscopic causes of the lag phase. J Theor Biol. 241 (4): 939–53. doi:10.1016/j.jtbi.2006.01.029. PMID 16524598.

102. Hecker M, Völker U (2001). General stress response of Bacillus subtilis and other bacteria. Adv Microb Physiol. Advances in Microbial Physiology. 44: 35–91. doi:10.1016/S0065-2911(01)44011-2. ISBN 978-0-12-027744-5. PMID 11407115.

103. Genomes Online Database. Genomes Online Database. Joint Genome Institute. Retrieved 14 Sep 2016.

104. Nakabachi A, Yamashita A, Toh H, Ishikawa H, Dunbar HE, Moran NA, Hattori M (2006). The 160-kilobase genome of the bacterial endosymbiont Carsonella. Science. 314 (5797): 267. doi:10.1126/science.1134196. PMID 17038615.

105. Pradella S, Hans A, Spröer C, Reichenbach H, Gerth K, Beyer S (2002). Characterisation, genome size and genetic manipulation of the myxobacterium Sorangium cellulosum So ce56. Arch Microbiol. 178 (6): 484–92. doi:10.1007/s00203-002-0479-2. PMID 12420170.

106. Belfort M, Reaban ME, Coetzee T, Dalgaard JZ (1 July 1995). Prokaryotic introns and inteins: a panoply of form and function. J. Bacteriol. 177 (14): 3897–903. PMC 177115Freely accessible. PMID 7608058.

107. Hinnebusch J, Tilly K (1993). Linear plasmids and chromosomes in bacteria. Mol Microbiol. 10 (5): 917–22. doi:10.1111/j.1365-2958.1993.tb00963.x. PMID 7934868.

108. Fraser, Claire M.; Casjens, Sherwood; Huang, Wai Mun; Sutton, Granger G.;

Clayton, Rebecca; Lathigra, Raju; White, Owen; Ketchum, Karen A.; Dodson, Robert (1997-12-11). Genomic sequence of a Lyme disease spirochaete, Borrelia burgdorferi. Nature. 390 (6660): 580–586. doi:10.1038/37551. ISSN 0028-0836.

109. The University of Waikato (March 25, 2014). Bacterial DNA – the role of plasmids. Themes — Bacteria in biotech. Biotechnology Learning Hub. Retrieved 2014-09-03.
110. Denamur E, Matic I (2006). Evolution of mutation rates in bacteria. Mol Microbiol. 60 (4): 820–7. doi:10.1111/j.1365-2958.2006.05150.x. PMID 16677295.
111. Wright BE (2004). Stress-directed adaptive mutations and evolution. Mol Microbiol. 52 (3): 643–50. doi:10.1111/j.1365-2958.2004.04012.x. PMID 15101972.
112. Chen I, Dubnau D (2004). DNA uptake during bacterial transformation. Nature Reviews Microbiology. 2 (3): 241–9. doi:10.1038/nrmicro844. PMID 15083159.
113. Solomon JM, Grossman AD (1996). Who's competent and when: regulation of natural genetic competence in bacteria. Trends Genet. 12 (4): 150–5. doi:10.1016/0168-9525(96)10014-7. PMID 8901420.
114. Akamatsu T, Taguchi H (2001). Incorporation of the whole chromosomal DNA in protoplast lysates into competent cells of Bacillus subtilis. Biosci. Biotechnol. Biochem. 65 (4): 823–9. doi:10.1271/bbb.65.823. PMID 11388459.
115. Saito Y, Taguchi H, Akamatsu T (2006). Fate of transforming bacterial genome following incorporation into competent cells of Bacillus subtilis: a continuous length of incorporated DNA. J. Biosci. Bioeng. 101 (3): 257–62. doi:10.1263/jbb.101.257. PMID 16716928.
116. Johnsborg O, Eldholm V, Håvarstein LS (2007). Natural genetic transformation: prevalence, mechanisms and function. Res. Microbiol. 158 (10): 767–78. doi:10.1016/j.resmic.2007.09.004. PMID 17997281.
117. Bernstein H, Bernstein C, Michod RE (2012). DNA repair as the primary adaptive function of sex in bacteria and eukaryotes. Chapter 1: pp. 1–49 in: DNA Repair: New Research, Sakura Kimura and Sora Shimizu (eds.). Nova Sci. Publ., Hauppauge, N.Y. ISBN 978-1-62100-808-8.
118. Michod RE, Bernstein H, Nedelcu AM (2008). Adaptive value of sex in microbial pathogens (PDF). Infect. Genet. Evol. 8 (3): 267–85. doi:10.1016/j.meegid.2008.01.002. PMID 18295550.
119. Davison J (1999). Genetic exchange between bacteria in the environment. Plasmid. 42 (2): 73–91. doi:10.1006/plas.1999.1421. PMID 10489325.
120. Hastings PJ, Rosenberg SM, Slack A (2004). Antibiotic-induced lateral transfer of antibiotic resistance. Trends Microbiol. 12 (9): 401–4. doi:10.1016/j.tim.2004.07.003. PMID 15337159.
121. Brüssow H, Canchaya C, Hardt WD (2004). Phages and the evolution of bacterial pathogens: from genomic rearrangements to lysogenic conversion. Microbiology and Molecular Biology Reviews. 68 (3): 560–602. doi:10.1128/MMBR.68.3.560-602.2004. PMC 515249Freely accessible. PMID 15353570.

122. Bickle TA, Krüger DH (1993). Biology of DNA restriction. Microbiol. Rev. 57 (2): 434–50. PMC 372918Freely accessible. PMID 8336674.

123. Barrangou R, Fremaux C, Deveau H, Richards M, Boyaval P, Moineau S, Romero DA, Horvath P (2007). CRISPR provides acquired resistance against viruses in prokaryotes. Science. 315 (5819): 1709–12. Bibcode:2007Sci...315.1709B. doi:10.1126/science.1138140. PMID 17379808.

124. Brouns SJ, Jore MM, Lundgren M, Westra ER, Slijkhuis RJ, Snijders AP, Dickman MJ, Makarova KS, Koonin EV, van der Oost J (2008). Small CRISPR RNAs guide antiviral defense in prokaryotes. Science. 321 (5891): 960–4. Bibcode:2008Sci...321..960B. doi:10.1126/science.1159689. PMID 18703739.

125. Dusenbery, David B. (1996). Life at Small Scale. Scientific American Library. ISBN 0-7167-5060-0.

126. Shapiro JA (1998). Thinking about bacterial populations as multicellular organisms (PDF). Annu. Rev. Microbiol. 52: 81–104. doi:10.1146/annurev.micro.52.1.81. PMID 9891794. Archived from the original (PDF) on 17 July 2011.

127. Costerton JW, Lewandowski Z, Caldwell DE, Korber DR, Lappin-Scott HM (1995). Microbial biofilms. Annu. Rev. Microbiol. 49: 711–45. doi:10.1146/annurev.mi.49.100195.003431. PMID 8561477.

128. Miller MB, Bassler BL (2001). Quorum sensing in bacteria. Annu. Rev. Microbiol. 55: 165–99. doi:10.1146/annurev.micro.55.1.165. PMID 11544353.

129. Bardy S, Ng S, Jarrell K (2003). Prokaryotic motility structures. Microbiology. 149 (Pt 2): 295–304. doi:10.1099/mic.0.25948-0. PMID 12624192.

130. Dusenbery, David B. (2009). Living at Micro Scale, p. 136. Harvard University Press, Cambridge, Mass. ISBN 978-0-674-03116-6.

131. Merz A, So M, Sheetz M (2000). Pilus retraction powers bacterial twitching motility. Nature. 407 (6800): 98–102. Bibcode:2000Natur.407...98M. doi:10.1038/35024105. PMID 10993081.

132. Dusenbery, David B. (2009). Living at Micro Scale, Chapter 13. Harvard University Press, Cambridge, Mass. ISBN 978-0-674-03116-6.

133. Macnab RM (1 December 1999). The bacterial flagellum: reversible rotary propellor and type III export apparatus. J. Bacteriol. 181 (23): 7149–53. PMC 103673Freely accessible. PMID 10572114.

134. Wu M, Roberts J, Kim S, Koch D, DeLisa M (2006). Collective bacterial dynamics revealed using a three-dimensional population-scale defocused particle tracking technique. Appl Environ Microbiol. 72 (7): 4987–94. doi:10.1128/AEM.00158-06. PMC 1489374Freely accessible. PMID 16820497.

135. Lux R, Shi W (2004). Chemotaxis-guided movements in bacteria. Crit Rev Oral Biol Med. 15 (4): 207–20. doi:10.1177/154411130401500404. PMID 15284186.

136. Schweinitzer T, Josenhans C (2010). Bacterial energy taxis: a global strategy?. Arch Microbiol. 192 (7): 507–20. doi:10.1007/s00203-010-0575-7. PMC 2886117Freely accessible. PMID 20411245.

137. Frankel R, Bazylinski D, Johnson M, Taylor B (1997). Magneto-aerotaxis in marine

coccoid bacteria. Biophys J. 73 (2): 994–1000. Bibcode:1997BpJ....73..994F. doi:10.1016/S0006-3495(97)78132-3. PMC 1180996Freely accessible. PMID 9251816.

138. Goldberg MB (2001). Actin-based motility of intracellular microbial pathogens. Microbiol Mol Biol Rev. 65 (4): 595–626, table of contents. doi:10.1128/MMBR.65.4.595-626.2001. PMC 99042Freely accessible. PMID 11729265.

139. Boucher Y, Douady CJ, Papke RT, Walsh DA, Boudreau ME, Nesbo CL, Case RJ, Doolittle WF (2003). Lateral gene transfer and the origins of prokaryotic groups. Annu Rev Genet. 37: 283–328. doi:10.1146/annurev.genet.37.050503.084247. PMID 14616063.

140. Olsen GJ, Woese CR, Overbeek R (1994). The winds of (evolutionary) change: breathing new life into microbiology. Journal of Bacteriology. 176 (1): 1–6. doi:10.2172/205047. PMC 205007Freely accessible. PMID 8282683.

141. IJSEM Home. Ijs.sgmjournals.org. 28 October 2011. Retrieved 4 November 2011.

142. Bergey's Manual Trust. Bergeys.org. Retrieved 4 November 2011.

143. Gupta R (2000). The natural evolutionary relationships among prokaryotes. Crit Rev Microbiol. 26 (2): 111–31. doi:10.1080/10408410091154219. PMID 10890353.

144. Doolittle RF (2005). Evolutionary aspects of whole-genome biology. Current Opinion in Structural Biology. 15 (3): 248–53. doi:10.1016/j.sbi.2005.04.001. PMID 15963888.

145. Cavalier-Smith T (2002). The neomuran origin of archaebacteria, the negibacterial root of the universal tree and bacterial megaclassification. Int J Syst Evol Microbiol. 52 (Pt 1): 7–76. doi:10.1099/00207713-52-1-7. PMID 11837318.

146. Ciccarelli FD, Doerks T, von Mering C, Creevey CJ, Snel B, Bork P (March 2006). Toward automatic reconstruction of a highly resolved tree of life. Science. 311 (5765): 1283–7. Bibcode:2006Sci...311.1283C. doi:10.1126/science.1123061. PMID 16513982.

147. Woods GL, Walker DH (1996). Detection of infection or infectious agents by use of cytologic and histologic stains. Clinical Microbiology Reviews. 9 (3): 382–404. PMC 172900Freely accessible. PMID 8809467.

148. Weinstein M (1994). Clinical importance of blood cultures. Clin Lab Med. 14 (1): 9–16. PMID 8181237.

149. Louie M, Louie L, Simor AE (8 August 2000). The role of DNA amplification technology in the diagnosis of infectious diseases. CMAJ. 163 (3): 301–9. PMC 80298Freely accessible. PMID 10951731.

150. Oliver J (2005). The viable but nonculturable state in bacteria. J Microbiol. 43 Spec No: 93–100. PMID 15765062. Archived from the original on 28 September 2007.

151. Euzéby JP (8 December 2011). Number of published names. List of Prokaryotic names with Standing in Nomenclature. Archived from the original on 19 January 2012. Retrieved 10 December 2011.

152. Curtis TP, Sloan WT, Scannell JW (2002). Estimating prokaryotic diversity and its limits. Proceedings of the National Academy of Sciences of the United States of America. 99 (16): 10494–9. Bibcode:2002PNAS...9910494C. doi:10.1073/pnas.142680199. PMC 124953Freely accessible. PMID 12097644.

153. Schloss PD, Handelsman J (2004). Status of the microbial census. Microbiology and Molecular Biology Reviews. 68 (4): 686–91. doi:10.1128/MMBR.68.4.686-691.2004. PMC 539005Freely accessible. PMID 15590780.

154. Martin MO (2002). Predatory prokaryotes: an emerging research opportunity. Journal of Microbiology and Biotechnology. 4 (5): 467–77. PMID 12432957.

155. Velicer GJ, Stredwick KL (2002). Experimental social evolution with Myxococcus xanthus. Antonie Van Leeuwenhoek. 81 (1–4): 155–64. doi:10.1023/A:1020546130033. PMID 12448714.

156. Gromov, BV. Electron Microscope Study of Parasitism by Bdellovibrio Chorellavorus Bacteria on Cells of the Green Alga Chorella Vulgaris. Tsitologiya 14.2 (1972): 256-60. Print.

157. Guerrero R, Pedros-Alio C, Esteve I, Mas J, Chase D, Margulis L (April 1986). Predatory prokaryotes: predation and primary consumption evolved in bacteria. Proceedings of the National Academy of Sciences of the United States of America. 83 (7): 2138–42. Bibcode:1986PNAS...83.2138G. doi:10.1073/pnas.83.7.2138. PMC 323246Freely accessible. PMID 11542073.

158. Velicer GJ, Mendes-Soares H (2009). Bacterial predators. Current Biology. 19 (2): R55–6. doi:10.1016/j.cub.2008.10.043. PMID 19174136.

159. Stams AJ, de Bok FA, Plugge CM, van Eekert MH, Dolfing J, Schraa G (2006). Exocellular electron transfer in anaerobic microbial communities. Environ Microbiol. 8 (3): 371–82. doi:10.1111/j.1462-2920.2006.00989.x. PMID 16478444.

160. Barea JM, Pozo MJ, Azcón R, Azcón-Aguilar C (2005). Microbial co-operation in the rhizosphere. J Exp Bot. 56 (417): 1761–78. doi:10.1093/jxb/eri197. PMID 15911555.

161. O'Hara AM, Shanahan F (2006). The gut flora as a forgotten organ. EMBO Reports. 7 (7): 688–93. doi:10.1038/sj.embor.7400731. PMC 1500832Freely accessible. PMID 16819463.

162. Zoetendal EG, Vaughan EE, de Vos WM (2006). A microbial world within us. Mol Microbiol. 59 (6): 1639–50. doi:10.1111/j.1365-2958.2006.05056.x. PMID 16553872.

163. Gorbach SL (1990). Lactic acid bacteria and human health. Annals of Medicine. 22 (1): 37–41. doi:10.3109/07853899009147239. PMID 2109988.

164. Salminen SJ, Gueimonde M, Isolauri E (1 May 2005). Probiotics that modify disease risk. J Nutr. 135 (5): 1294–8. PMID 15867327.

165. Fish DN (2002). Optimal antimicrobial therapy for sepsis. Am J Health Syst Pharm. 59 Suppl 1: S13–9. PMID 11885408.

166. Belland RJ, Ouellette SP, Gieffers J, Byrne GI (2004). Chlamydia pneumoniae and atherosclerosis. Cell Microbiol. 6 (2): 117–27. doi:10.1046/j.1462-

5822.2003.00352.x. PMID 14706098.

167. Heise ER (1982). Diseases associated with immunosuppression. Environmental Health Perspectives. 43: 9–19. doi:10.2307/3429162. JSTOR 3429162. PMC 1568899Freely accessible. PMID 7037390.
168. Saiman L (2004). Microbiology of early CF lung disease. Paediatric Respiratory Reviews. 5 Suppl A: S367–9. doi:10.1016/S1526-0542(04)90065-6. PMID 14980298.
169. Fisher B, Harvey RP, Champe PC (2007). Lippincott's Illustrated Reviews: Microbiology (Lippincott's Illustrated Reviews Series). Hagerstwon, MD: Lippincott Williams & Wilkins. pp. Chapter 33, pages 367–392. ISBN 0-7817-8215-5.
170. LEF.org > Bacterial Infections Updated: 19 January 2006. Retrieved on 11 April 2009
171. Yonath A, Bashan A (2004). Ribosomal crystallography: initiation, peptide bond formation, and amino acid polymerization are hampered by antibiotics. Annu Rev Microbiol. 58: 233–51. doi:10.1146/annurev.micro.58.030603.123822. PMID 15487937.
172. Khachatourians GG (1998). Agricultural use of antibiotics and the evolution and transfer of antibiotic-resistant bacteria. CMAJ. 159 (9): 1129–36. PMC 1229782Freely accessible. PMID 9835883.
173. Johnson ME, Lucey JA (2006). Major technological advances and trends in cheese. J Dairy Sci. 89 (4): 1174–8. doi:10.3168/jds.S0022-0302(06)72186-5. PMID 16537950.
174. Hagedorn S, Kaphammer B (1994). Microbial biocatalysis in the generation of flavor and fragrance chemicals. Annu. Rev. Microbiol. 48: 773–800. doi:10.1146/annurev.mi.48.100194.004013. PMID 7826026.
175. Cohen Y (2002). Bioremediation of oil by marine microbial mats. Int Microbiol. 5 (4): 189–93. doi:10.1007/s10123-002-0089-5. PMID 12497184.
176. Neves LC, Miyamura TT, Moraes DA, Penna TC, Converti A (2006). Biofiltration methods for the removal of phenolic residues. Appl. Biochem. Biotechnol. 129–132: 130–52. doi:10.1385/ABAB:129:1:130. PMID 16915636.
177. Liese A, Filho MV (1999). Production of fine chemicals using biocatalysis. Current Opinion in Biotechnology. 10 (6): 595–603. doi:10.1016/S0958-1669(99)00040-3. PMID 10600695.
178. Aronson AI, Shai Y (2001). Why Bacillus thuringiensis insecticidal toxins are so effective: unique features of their mode of action. FEMS Microbiol. Lett. 195 (1): 1–8. doi:10.1111/j.1574-6968.2001.tb10489.x. PMID 11166987.
179. Bozsik A (2006). Susceptibility of adult Coccinella septempunctata (Coleoptera: Coccinellidae) to insecticides with different modes of action. Pest Manag Sci. 62 (7): 651–4. doi:10.1002/ps.1221. PMID 16649191.
180. Chattopadhyay A, Bhatnagar NB, Bhatnagar R (2004). Bacterial insecticidal toxins. Crit Rev Microbiol. 30 (1): 33–54. doi:10.1080/10408410490270712.

PMID 15116762.

181. Serres MH, Gopal S, Nahum LA, Liang P, Gaasterland T, Riley M (2001). A functional update of the Escherichia coli K-12 genome. Genome Biology. 2 (9): research0035.1–research0035.7. doi:10.1186/gb-2001-2-9-research0035. PMC 56896Freely accessible. PMID 11574054.

182. Almaas E, Kovács B, Vicsek T, Oltvai ZN, Barabási AL (2004). Global organization of metabolic fluxes in the bacterium Escherichia coli. Nature. 427 (6977): 839–43. arXiv:q-bio/0403001Freely accessible. Bibcode:2004Natur.427..839A. doi:10.1038/nature02289. PMID 14985762.

183. Reed JL, Vo TD, Schilling CH, Palsson BO (2003). An expanded genome-scale model of Escherichia coli K-12 (iJR904 GSM/GPR). Genome Biol. 4 (9): R54. doi:10.1186/gb-2003-4-9-r54. PMC 193654Freely accessible. PMID 12952533.

184. Walsh G (2005). Therapeutic insulins and their large-scale manufacture. Appl Microbiol Biotechnol. 67 (2): 151–9. doi:10.1007/s00253-004-1809-x. PMID 15580495.

185. Graumann K, Premstaller A (2006). Manufacturing of recombinant therapeutic proteins in microbial systems. Biotechnol J. 1 (2): 164–86. doi:10.1002/biot.200500051. PMID 16892246.

186. Porter JR (1976). Antony van Leeuwenhoek: tercentenary of his discovery of bacteria. Bacteriological Reviews. 40 (2): 260–9. PMC 413956Freely accessible. PMID 786250.

187. van Leeuwenhoek A (1684). An abstract of a letter from Mr. Anthony Leevvenhoek at Delft, dated Sep. 17, 1683, Containing Some Microscopical Observations, about Animals in the Scurf of the Teeth, the Substance Call'd Worms in the Nose, the Cuticula Consisting of Scales. Philosophical Transactions (1683–1775). 14 (155–166): 568–574. doi:10.1098/rstl.1684.0030.

188. van Leeuwenhoek A (1700). Part of a Letter from Mr Antony van Leeuwenhoek, concerning the Worms in Sheeps Livers, Gnats, and Animalcula in the Excrements of Frogs. Philosophical Transactions (1683–1775). 22 (260–276): 509–518. doi:10.1098/rstl.1700.0013.

189. van Leeuwenhoek A (1702). Part of a Letter from Mr Antony van Leeuwenhoek, F. R. S. concerning Green Weeds Growing in Water, and Some Animalcula Found about Them. Philosophical Transactions (1683–1775). 23 (277–288): 1304–11. doi:10.1098/rstl.1702.0042.

190. Asimov, Isaac (1982), Asimov's Biographical Encyclopedia of Science and Technology, 2nd edition, Garden City, New York: Doubleday and Company, pg 143.

191. Ehrenberg's Symbolae Physioe. Animalia evertebrata. Decas prima. Berlin, 1828.

192. Breed RS, Conn HJ (1936). The Status of the Generic Term Bacterium Ehrenberg 1828. Journal of Bacteriology. 31 (5): 517–518. PMC 543738Freely accessible. PMID 16559906.

193. EHRENBERG (C.G.): Dritter Beitrag zur Erkenntniss grosser Organisation in der

Richtung des kleinsten Raumes. Physikalische Abhandlungen der Koeniglichen Akademie der Wissenschaften zu Berlin aus den Jahren 1833–1835, 1835, pp. 143–336.

194. Pasteur's Papers on the Germ Theory. LSU Law Center's Medical and Public Health Law Site, Historic Public Health Articles. Archived from the original on 18 December 2006. Retrieved 23 November 2006.

195. The Nobel Prize in Physiology or Medicine 1905. Nobelprize.org. Archived from the original on 10 December 2006. Retrieved 22 November 2006.

196. O'Brien SJ, Goedert JJ (1996). HIV causes AIDS: Koch's postulates fulfilled. Current Opinion in Immunology. 8 (5): 613–8. doi:10.1016/S0952-7915(96)80075-6. PMID 8902385.

197. Thurston AJ (2000). Of blood, inflammation and gunshot wounds: the history of the control of sepsis. Aust N Z J Surg. 70 (12): 855–61. doi:10.1046/j.1440-1622.2000.01983.x. PMID 11167573.

198. Schwartz RS (2004). Paul Ehrlich's magic bullets. N Engl J Med. 350 (11): 1079–80. doi:10.1056/NEJMp048021. PMID 15014180.

199. Biography of Paul Ehrlich. Nobelprize.org. Archived from the original on 28 November 2006. Retrieved 26 November 2006.

CHAPTER 6

Organic Compounds

Chapter Contents

An **organic compound** is virtually any chemical compound that contains carbon, although a consensus definition remains elusive and likely arbitrary. Organic compounds are rare terrestrially, but of central importance because all known life is based on organic compounds. The most basic petrochemicals are considered the building blocks of organic chemistry.

INTRODUCTION

For historical reasons discussed below, a few types of carbon-containing compounds, such as carbides, carbonates, simple oxides of carbon (for example, CO and CO_2), and cyanides are considered inorganic. The distinction between *organic and inorganic* carbon compounds, while "useful in organizing the vast subject of chemistry... is somewhat arbitrary".

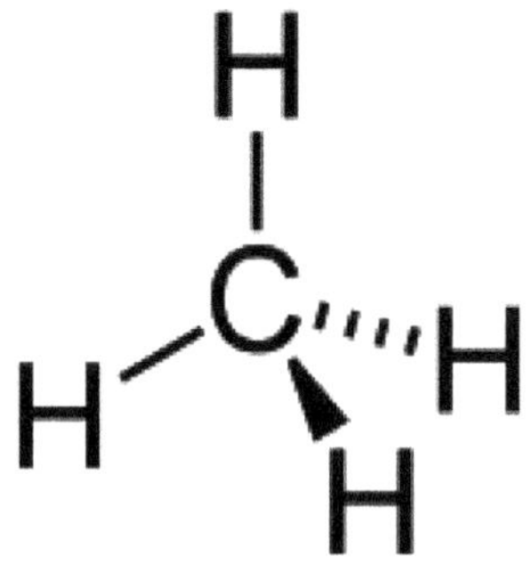

Methane, CH_4; it is one of the simplest organic compounds.

Organic chemistry is the science concerned with all aspects of organic compounds. Organic synthesis is the methodology of their preparation.

HISTORY

Vitalism

The word *organic* is historical, dating to the 1st century. For many centuries, Western alchemists believed in vitalism. This is the theory that certain compounds could be synthesized only from their classical elements—earth, water, air, and fire—by the action of a "life-force" (*vis vitalis*) that only organisms possessed. Vitalism taught that these "organic" compounds were fundamentally different from the "inorganic" compounds that could be obtained from the elements by chemical manipulation.

Vitalism survived for a while even after the rise of modern atomic theory and the replacement of the Aristotelian elements by those we know today. It first came under question in 1824, when Friedrich Wöhler synthesized oxalic acid, a compound known to occur only in living organisms, from cyanogen. A more decisive experiment was Wöhler's 1828 synthesis of urea from the inorganic salts potassium cyanate and ammonium sulfate. Urea had long been considered an "organic" compound, as it was known to occur only in the urine of living organisms. Wöhler's experiments were followed by many others, in which increasingly complex "organic" substances were produced from "inorganic" ones without the involvement of any living organism.

Modern Classification

Even though vitalism has been discredited, scientific nomenclature retains the distinction between *organic* and *inorganic* compounds. The modern meaning of *organic compound* is any compound that contains a significant amount of carbon—even though many of the organic compounds known today have no connection to any substance found in living organisms.

The organic compound L-isoleucine molecule presents some features typical of organic compounds: carbon–carbon bonds, carbon–hydrogen bonds, as well as covalent bonds between carbon to oxygen and to nitrogen.

Still, even the broadest definition (of "carbon-containing molecules" as organic) requires excluding alloys that contain carbon, including steel. Other 'excluded' materials are: compounds such as carbonates and carbonyls, simple oxides of carbon, simple carbon halides and sulfides, the allotropes of carbon, and cyanides not containing the −C≡N functional group—all which are considered inorganic.

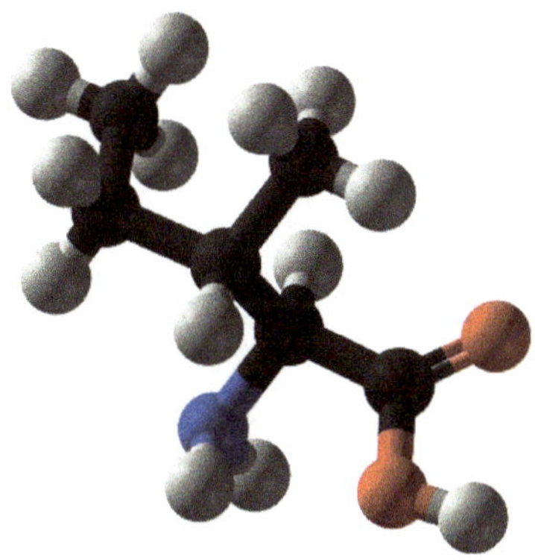

The L-isoleucine molecule, $C_6H_{13}NO_2$, showing features typical of organic compounds. Carbon atoms are in black, hydrogens gray, oxygens red, and nitrogen blue.

The "C-H" definition excludes compounds that are (historically and practically) considered organic. Neither urea nor oxalic acid is organic by this definition, yet they were two key compounds in the vitalism debate. The IUPAC Blue Book on organic nomenclature specifically mentions urea and oxalic acid. Other compounds lacking C-H bonds but traditionally considered organic include benzenehexol, mesoxalic acid, and carbon tetrachloride. Mellitic acid, which contains no C-H bonds, is considered a possible organic substance in Martian soil.

The "C-H bond-only" rule also leads to somewhat arbitrary divisions in sets of carbon-fluorine compounds. For example, CF_4 would be considered by this rule to be "inorganic", whereas CF_3H would be organic.

CLASSIFICATION

Organic compounds may be classified in a variety of ways. One major distinction is between natural and synthetic compounds. Organic compounds can also be classified or subdivided by the presence of heteroatoms, e.g., organometallic compounds, which feature bonds between carbon and a metal, and organophosphorus compounds, which feature bonds between carbon and a phosphorus.

Another distinction, based on the size of organic compounds, distinguishes between small molecules and polymers.

Natural Compounds

Natural compounds refer to those that are produced by plants or animals. Many of these are still extracted from natural sources because they would be more expensive to produce artificially. Examples include most sugars, some alkaloids and terpenoids, certain nutrients such as vitamin B_{12}, and, in general, those natural products with large or stereoisometrically complicated molecules present in reasonable concentrations in living organisms.

Further compounds of prime importance in biochemistry are antigens, carbohydrates, enzymes, hormones, lipids and fatty acids, neurotransmitters, nucleic acids, proteins, peptides and amino acids, lectins, vitamins, and fats and oils.

Synthetic Compounds

Compounds that are prepared by reaction of other compounds are known as "synthetic". They may be either compounds that already are found in plants or animals or those that do not occur naturally.

Most polymers (a category that includes all plastics and rubbers), are organic synthetic or semi-synthetic compounds.

Biotechnology

Many organic compounds—two examples are ethanol and insulin—are manufactured industrially using organisms such as bacteria and yeast. Typically, the DNA of an organism is altered to express compounds not ordinarily produced by the organism. Many such biotechnology-engineered compounds did not previously exist in nature.

DATABASES

- The *CAS* database is the most comprehensive repository for data on organic compounds. The search tool *SciFinder* is offered.
- The *Beilstein database* contains information on 9.8 million substances, covers the scientific literature from 1771 to the present, and is today accessible via Reaxys. Structures and a large diversity of physical and chemical properties is available for each substance, with reference to original literature.
- *PubChem* contains 18.4 million entries on compounds and especially covers the field of medicinal chemistry.

A great number of more specialized databases exist for diverse branches of organic chemistry.

STRUCTURE DETERMINATION

The main tools are proton and carbon-13 NMR spectroscopy, IR Spectroscopy, Mass spectrometry, UV/Vis Spectroscopy and X-ray crystallography.

Organic Chemistry

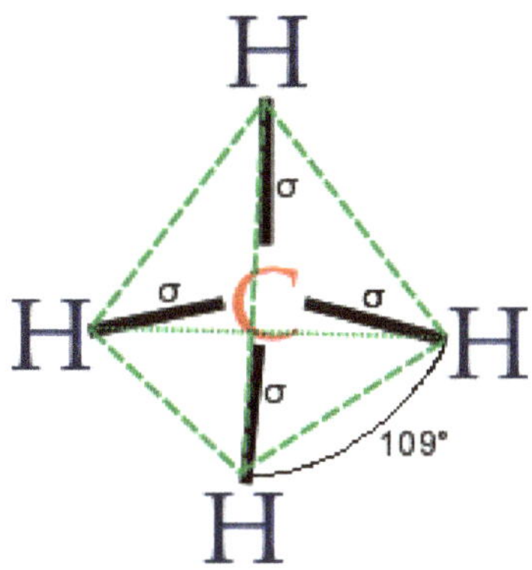

Methane, CH_4; the line-angle structural formula shows four carbon-hydrogen single bonds (σ, in black), and the typical 3D shape of tetrahedral molecules, with ~109° interior bond angles (in dashed-green).

Organic chemistry is a chemistry subdiscipline involving the scientific study of the structure, properties, and reactions of organic compounds and organic materials, i.e., matter in its various forms that contain carbon atoms.Study of structure includes many physical and chemical methods to determine the chemical composition and the chemical constitution of organic compounds and materials. Study of properties includes both physical properties and chemical properties, and uses similar methods as well as methods to evaluate chemical reactivity, with the aim to understand the behavior of the organic matter in its pure form (when possible), but also in solutions, mixtures, and fabricated forms. The study of organic reactions includes probing their scope through use in preparation of target compounds (e.g., natural products, drugs, polymers, etc.) by chemical synthesis, as well as the focused study of the reactivities of individual organic molecules, both in the laboratory and via theoretical (in silico) study.

The range of chemicals studied in organic chemistry include hydrocarbons (compounds containing only carbon and hydrogen), as well as myriad compositions based always on carbon, but also containing other elements,especially oxygen, nitrogen, sulfur, phosphorus (these, included in many organic chemicals in biology) and the radiostable elements of the halogens.

In the modern era, the range extends further into the periodic table, with main group elements, including:

- Group 1 and 2 organometallic compounds, i.e., involving alkali (e.g., lithium, sodium, and potassium) or alkaline earth metals (e.g., magnesium)
- Metalloids (e.g., boron and silicon) or other metals (e.g., aluminium and tin)

In addition, much modern research focuses on organic chemistry involving further organometallics, including the lanthanides, but especially the transition metals; (e.g., zinc, copper, palladium, nickel, cobalt, titanium and chromium)

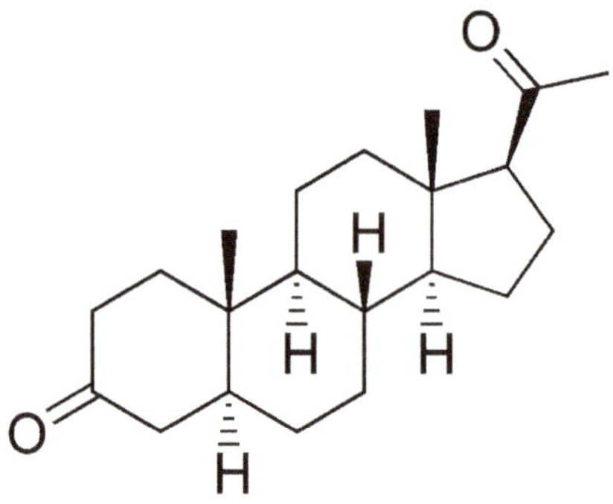

Line-angle representation.

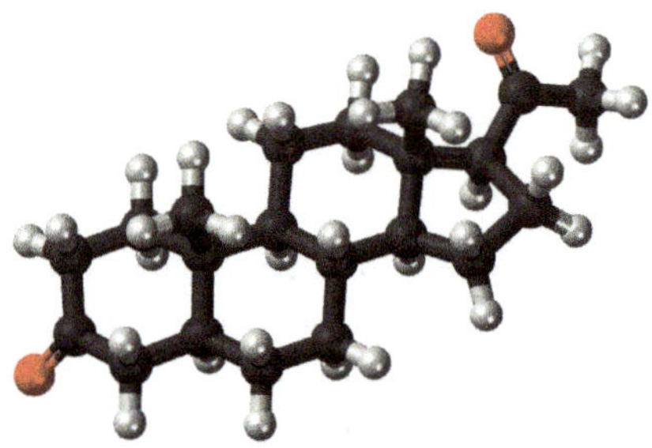

Ball-and-stick representation.

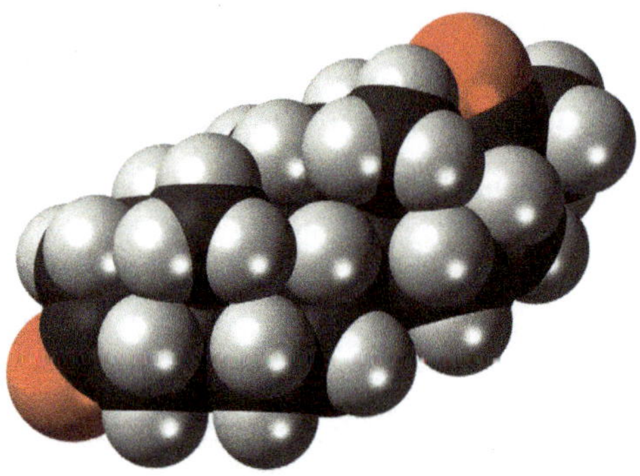

Space-filling representation.

Three representations of an organic compound, 5α-Dihydroprogesterone (5α-DHP), a steroid hormone. For molecules showing color, the carbon atoms are in black, hydrogens in gray, and oxygens in red. In the line angle representation, carbon atoms are implied at every terminus of a line and vertex of multiple lines, and hydrogen atoms are implied to fill the remaining needed valences (up to 4).

Finally, organic compounds form the basis of all earthly life and constitute a significant part of human endeavors in chemistry. The bonding patterns open to carbon, with its valence of four—formal single, double, and triple bonds, as well as various structures with delocalized electrons—make the array of organic compounds structurally diverse, and their range of applications enormous. They either form the basis of, or are important constituents of, many commercial products including pharmaceuticals; petrochemicals and products made from them (including lubricants, solvents, etc.); plastics; fuels and explosives; etc. As indicated, the study of organic chemistry overlaps with organometallic chemistry and biochemistry, but also with medicinal chemistry, polymer chemistry, as well as many aspects of materials science.

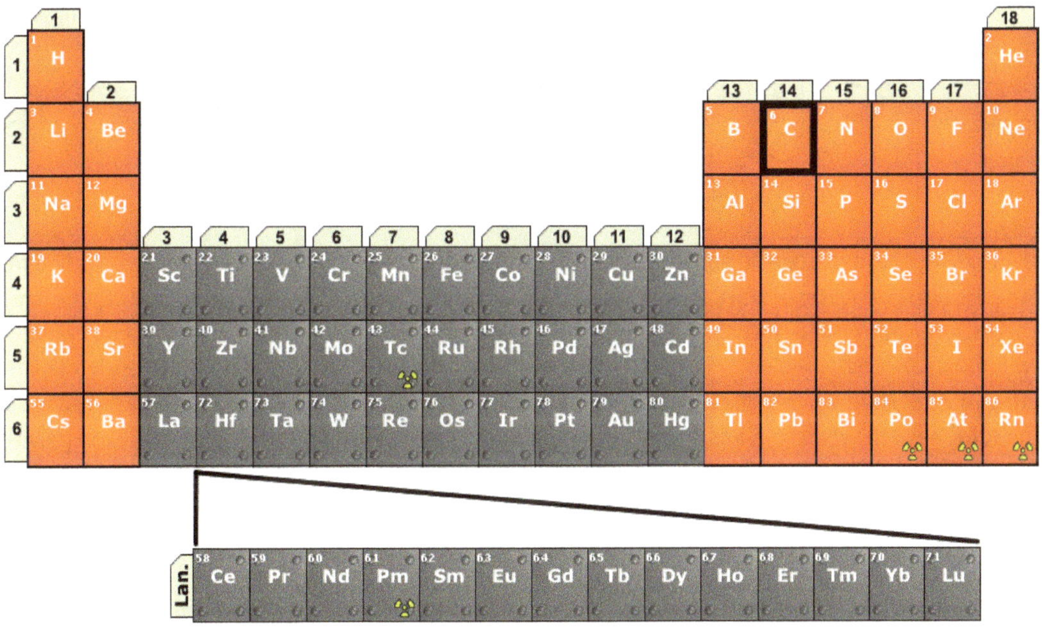

Periodic table of elements of interest in organic chemistry. The table illustrates all elements of current interest in modern organic and organometallic chemistry, indicating main group elements in orange, and transition metals and lanthanides (Lan) in grey.

Friedrich Wöhler.

Before the nineteenth century, chemists generally believed that compounds obtained from living organisms were endowed with a vital force that distinguished them from inorganic compounds. According to the concept of vitalism (vital force theory), organic matter was endowed with a "vital force". During the first half of the nineteenth century, some of the first systematic studies of organic compounds were reported. Around 1816 Michel Chevreul started a study of soaps made from various fats and alkalis. He separated the different acids that, in combination with the alkali, produced the soap. Since these were all individual compounds, he demonstrated that it was possible to make a chemical change in various fats (which traditionally come from organic sources), producing new compounds, without "vital force". In 1828 Friedrich Wöhler produced the *organic* chemical urea (carbamide), a constituent of urine, from *inorganic* starting materials (the salts potassium cyanate and ammonium sulfate), in what is now called the Wöhler synthesis. Although Wöhler himself was cautious about claiming he had disproved vitalism, this was the first time a substance thought to be organic was synthesized in the laboratory without biological (organic) starting materials. The event is now generally accepted as indeed disproving the doctrine of vitalism.

In 1856 William Henry Perkin, while trying to manufacture quinine accidentally produced the organic dye now known as Perkin's mauve. His discovery, made widely known through its financial success, greatly increased interest in organic chemistry.

A crucial breakthrough for organic chemistry was the concept of chemical structure, developed independently in 1858 by both Friedrich August Kekulé and Archibald Scott Couper. Both researchers suggested that tetravalent carbon atoms could link to each other to form a carbon lattice, and that the detailed patterns of atomic bonding could be discerned by skillful interpretations of appropriate chemical reactions.

The era of the pharmaceutical industry began in the last decade of the 19th century when the manufacturing of acetylsalicylic acid—more commonly referred to as aspirin—in Germany was started by Bayer. By 1910 Paul Ehrlich and his laboratory group began developing arsenic-based arsphenamine, (Salvarsan), as the first effective medicinal treatment of syphilis, and thereby initiated the medical practice of chemotherapy. Ehrlich popularized the concepts of "magic bullet" drugs and of systematically improving drug therapies. His laboratory made decisive contributions to developing antiserum for diphtheria and standardizing therapeutic serums.

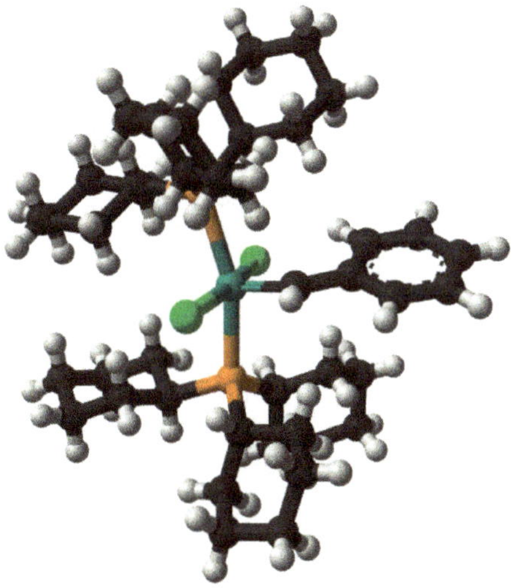

An example of an organometallic molecule, a catalyst called Grubbs' catalyst. Its formula is often given as $RuCl_2(PCy_3)_2(=CHPh)$, where the ball-and-stick model is based on X-ray crystallography. The single metal atom ruthenium (Ru), (in turquoise), is at the very center of the structure; two chlorines (green), are bonded to the ruthenium atom—carbon atoms are black, hydrogens gray-white, and phosphorus orange. A phosphorus-ligand bond, tricyclohexyl phosphine, PCy, is below center; (another PCy ligand appears at the top of the image where its rings are obscuring one another). The ring group projecting to the right, an alkylidene, contains a metal-carbon double bond to ruthenium.

Early examples of organic reactions and applications were often found because of a combination of luck and preparation for unexpected observations. The latter half of the 19th century however witnessed systematic studies of organic compounds. The development of synthetic indigo is illustrative. The production of indigo from plant sources dropped from 19,000 tons in 1897 to 1,000 tons by 1914 thanks to the synthetic methods developed by Adolf von Baeyer. In 2002, 17,000 tons of synthetic indigo were produced from petrochemicals.

In the early part of the 20th Century, polymers and enzymes were shown to be large organic molecules, and petroleum was shown to be of biological origin.

The multiple-step synthesis of complex organic compounds is called total synthesis. Total synthesis of complex natural compounds increased in complexity to glucose and terpineol. For example, cholesterol-related compounds have opened ways to synthesize complex human hormones and their modified derivatives. Since the start of the 20th century, complexity of total syntheses has been increased to include molecules of high complexity such as lysergic acid and vitamin B_{12}.

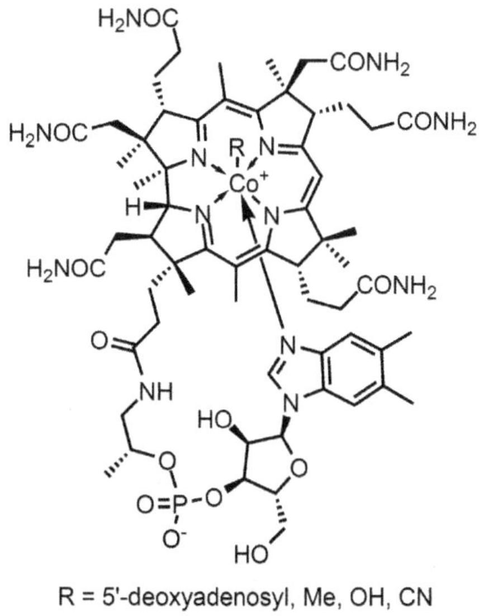

The total synthesis of vitamin B_{12} marked a major achievement in organic chemistry.

The discovery of petroleum and the development of the petrochemical industry spurred the development of organic chemistry. Converting individual petroleum compounds into different *types* of compounds by various chemical processes led to organic reactions enabling a broad range of industrial and commercial products including, among (many) others: plastics, synthetic rubber, organic adhesives, and various property-modifying petroleum additives and catalysts. The latter includes the Grubbs' catalyst, which mediates the important organic reactions called olefin metathesis. In 2005 Robert Grubbs shared the Nobel prize in chemistry with Richard R. Schrock and Yves Chauvin for their work on olefin metathesis using the Grubbs' catalyst.

The majority of chemical compounds occurring in biological organisms are in fact carbon compounds, so the association between organic chemistry and biochemistry is so close that biochemistry might be regarded as in essence a branch of organic chemistry. Although the history of biochemistry might be taken to span some four centuries, fundamental understanding of the field only began to develop in the late 19th century and the actual term *biochemistry* was coined around the start of 20th century. Research in the field increased throughout the twentieth century, without any indication of slackening in the rate of increase, as may be verified by inspection of abstraction and indexing services such as BIOSIS Previews and Biological Abstracts, which began in the 1920s as a single annual volume, but has grown so drastically that by the end of the 20th century it was only available to the everyday user as an online electronic database.

Characterization

Since organic compounds often exist as mixtures, a variety of techniques have also been developed to assess purity, especially important being chromatography techniques such as HPLC and gas chromatography. Traditional methods of separation include distillation, crystallization, and solvent extraction.

Organic compounds were traditionally characterized by a variety of chemical tests, called "wet methods", but such tests have been largely displaced by spectroscopic or other computer-intensive methods of analysis. Listed in approximate order of utility, the chief analytical methods are:

- Nuclear magnetic resonance (NMR) spectroscopy is the most commonly used technique, often permitting complete assignment of atom connectivity and even stereochemistry using correlation spectroscopy. The principal constituent atoms of organic chemistry – hydrogen and carbon – exist naturally with NMR-responsive isotopes, respectively ^{1}H and ^{13}C.
- Elemental analysis: A destructive method used to determine the elemental composition of a molecule. See also mass spectrometry, below.
- Mass spectrometry indicates the molecular weight of a compound and, from the fragmentation patterns, its structure. High resolution mass spectrometry can usually identify the exact formula of a compound and is used in lieu of elemental analysis. In former times, mass spectrometry was restricted to neutral molecules exhibiting some volatility, but advanced ionization techniques allow one to obtain the "mass spec" of virtually any organic compound.
- Crystallography can be useful for determining molecular geometry when a single crystal of the material is available and the crystal is representative of the sample. Highly automated software allows a structure to be determined within hours of obtaining a suitable crystal.

Traditional spectroscopic methods such as infrared spectroscopy, optical rotation, UV/VIS spectroscopy provide relatively nonspecific structural information but remain in use for specific classes of compounds. Traditionally refractive index and density were also important for substance identification.

Properties

Physical properties of organic compounds typically of interest include both quantitative and qualitative features. Quantitative information includes melting point, boiling point, and index of refraction. Qualitative properties include odor, consistency, solubility, and color.

Melting and Boiling Properties

Organic compounds typically melt and many boil. In contrast, while inorganic materials generally can be melted, many do not boil, tending instead to degrade. In earlier times, the melting point (m.p.) and boiling point (b.p.) provided crucial information on the purity and identity of organic compounds. The melting and boiling points correlate with the polarity of the molecules and their molecular weight. Some organic compounds, especially symmetrical ones, sublime, that is they evaporate without melting. A well-known example of a sublimable organic compound is para-dichlorobenzene, the odiferous constituent of modern mothballs. Organic compounds are usually not very stable at temperatures above 300 °C, although some exceptions exist.

Solubility

Neutral organic compounds tend to be hydrophobic; that is, they are less soluble in water than in organic solvents. Exceptions include organic compounds that contain ionizable (which can be converted in ions) groups as well as low molecular weight alcohols, amines, and carboxylic acids where hydrogen bonding occurs. Organic compounds tend to dissolve in organic solvents. Solvents can be either pure substances like ether or ethyl alcohol, or mixtures, such as the paraffinic solvents such as the various petroleum ethers and white spirits, or the range of pure or mixed aromatic solvents obtained from petroleum or tar fractions by physical separation or by chemical conversion. Solubility in the different solvents depends upon the solvent type and on the functional groups if present in the solution.

Solid State Properties

Various specialized properties of molecular crystals and organic polymers with conjugated systems are of interest depending on applications, e.g. thermo-mechanical and electro-mechanical such as piezoelectricity, electrical conductivity (see conductive polymers and organic semiconductors), and electro-optical (e.g. non-linear optics) properties. For historical reasons, such properties are mainly the subjects of the areas of polymer science and materials science.

Nomenclature

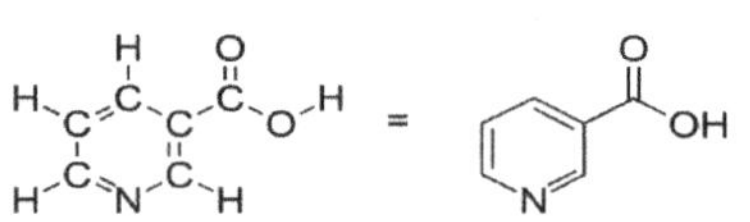

= C_5H_4N-3-CO_2H = pyridine-3-carboxylic acid = niacin = vitamin B_3

Various names and depictions for one organic compound.

The names of organic compounds are either systematic, following logically from a set of rules, or nonsystematic, following various traditions. Systematic nomenclature is stipulated by specifications from IUPAC. Systematic nomenclature starts with the name for a parent structure within the molecule of interest. This parent name is then modified by prefixes, suffixes, and numbers to unambiguously convey the structure. Given that millions of organic compounds are known, rigorous use of systematic names can be cumbersome. Thus, IUPAC recommendations are more closely followed for simple compounds, but not complex molecules. To use the systematic naming, one must know the structures and names of the parent structures. Parent structures include unsubstituted hydrocarbons, heterocycles, and monofunctionalized derivatives thereof.

Nonsystematic nomenclature is simpler and unambiguous, at least to organic chemists. Nonsystematic names do not indicate the structure of the compound. They are common for complex molecules, which includes most natural products. Thus, the informally named lysergic acid diethylamide is systematically named (6a*R*,9*R*)-*N*,*N*-diethyl-7-methyl-4,6,6a,7,8,9-hexahydroindolo-[4,3-*fg*] quinoline-9-carboxamide.

With the increased use of computing, other naming methods have evolved that are intended to be interpreted by machines. Two popular formats are SMILES and InChI.

Structural Drawings

Organic molecules are described more commonly by drawings or structural formulas, combinations of drawings and chemical symbols. The line-angle formula is simple and unambiguous. In this system, the endpoints and intersections of each line represent one carbon, and hydrogen atoms can either be notated explicitly or assumed to be present as implied by tetravalent carbon. The depiction of organic compounds with drawings is greatly simplified by the fact that carbon in almost all organic compounds has four bonds, nitrogen three, oxygen two, and hydrogen one.

Classification of Organic Compounds

Functional Groups

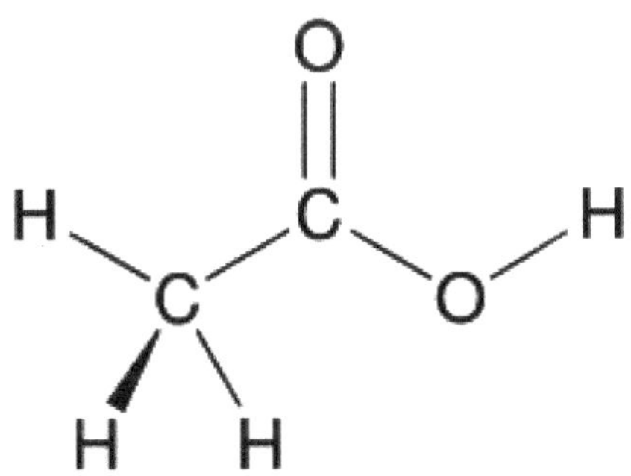

The family of carboxylic acids contains a carboxyl (-COOH) functional group. Acetic acid, shown here, is an example.

The concept of functional groups is central in organic chemistry, both as a means to classify structures and for predicting properties. A functional group is a molecular module, and the reactivity of that functional group is assumed, within limits, to be the same in a variety of molecules. Functional groups can have decisive influence on the chemical and physical properties of organic compounds. Molecules are classified on the basis of their functional groups. Alcohols, for example, all have the subunit C-O-H. All alcohols tend to be somewhat hydrophilic, usually form esters, and usually can be converted to the corresponding halides. Most functional groups feature heteroatoms (atoms other than C and H). Organic compounds are classified according to functional groups, alcohols, carboxylic acids, amines, etc.

Aliphatic Compounds

The aliphatic hydrocarbons are subdivided into three groups of homologous series according to their state of saturation:

- paraffins, which are alkanes without any double or triple bonds,
- olefins or alkenes which contain one or more double bonds, i.e. di-olefins (dienes) or poly-olefins.
- alkynes, which have one or more triple bonds.

The rest of the group is classed according to the functional groups present. Such compounds can be "straight-chain", branched-chain or cyclic. The degree of branching affects characteristics, such as the octane number or cetane number in petroleum chemistry.

Both saturated (alicyclic) compounds and unsaturated compounds exist as cyclic derivatives. The most stable rings contain five or six carbon atoms, but large rings (macrocycles) and smaller rings are common. The smallest cycloalkane family is the three-membered cyclopropane ($(CH_2)_3$). Saturated cyclic compounds contain single bonds only, whereas aromatic rings have an alternating (or conjugated) double bond. Cycloalkanes do not contain multiple bonds, whereas the cycloalkenes and the cycloalkynes do.

Aromatic Compounds

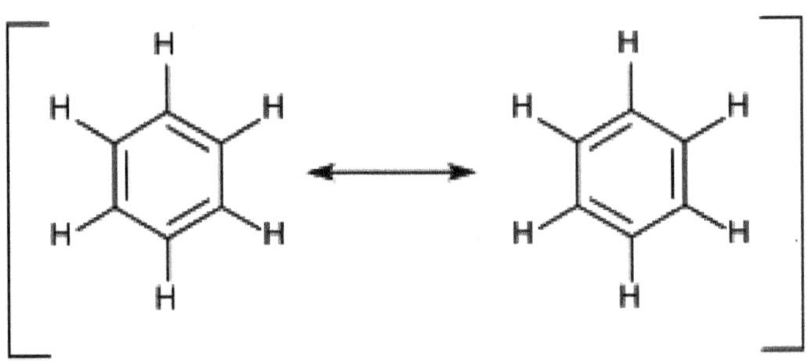

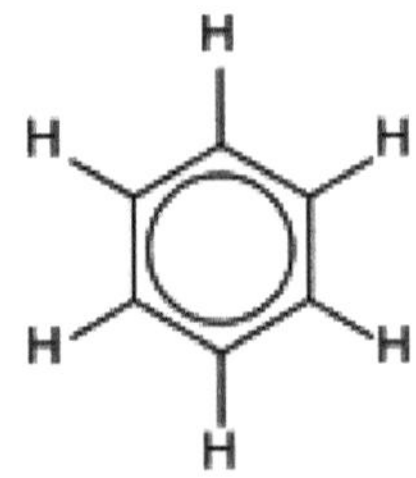

Benzene is one of the best-known aromatic compounds as it is one of the simplest and most stable aromatics.

Aromatic hydrocarbons contain conjugated double bonds. This means that every carbon atom in the ring is sp2 hybridized, allowing for added stability. The most important example is benzene, the structure of which was formulated by Kekulé who first proposed the delocalization or resonance principle for explaining its structure. For "conventional" cyclic compounds, aromaticity is conferred by the presence of 4n + 2 delocalized pi electrons, where n is an integer. Particular instability (antiaromaticity) is conferred by the presence of 4n conjugated pi electrons.

Heterocyclic Compounds

The characteristics of the cyclic hydrocarbons are again altered if heteroatoms are present, which can exist as either substituents attached externally to the ring (exocyclic) or as a member of the ring itself (endocyclic). In the case of the latter, the ring is termed a heterocycle. Pyridine and furan are examples of aromatic heterocycles while piperidine and tetrahydrofuran are the corresponding alicyclic heterocycles. The heteroatom of heterocyclic molecules is generally oxygen, sulfur, or nitrogen, with the latter being particularly common in biochemical systems.

Heterocycles are commonly found in a wide range of products including aniline dyes and medicines. Additionally, they are prevalent in a wide range of biochemical compounds such as alkaloids, vitamins, steroids, and nucleic acids (e.g. DNA, RNA).

Rings can fuse with other rings on an edge to give polycyclic compounds. The purine nucleoside bases are notable polycyclic aromatic heterocycles. Rings can also fuse on a "corner" such that one atom (almost always carbon) has two bonds going to one ring and two to another. Such compounds are termed spiro and are important in a number of natural products.

Polymers

This swimming board is made of polystyrene, an example of a polymer.

One important property of carbon is that it readily forms chains, or networks, that are linked by carbon-carbon (carbon-to-carbon) bonds. The linking process is called polymerization, while the chains, or networks, are called polymers. The source compound is called a monomer.

Two main groups of polymers exist: synthetic polymers and biopolymers. Synthetic polymers are artificially manufactured, and are commonly referred to as industrial polymers. Biopolymers occur within a respectfully natural environment, or without human intervention.

Since the invention of the first synthetic polymer product, bakelite, synthetic polymer products have frequently been invented.

Common synthetic organic polymers are polyethylene (polythene), polypropylene, nylon, teflon (PTFE), polystyrene, polyesters, polymethylmethacrylate (called perspex and plexiglas), and polyvinylchloride (PVC).

Both synthetic and natural rubber are polymers.

Varieties of each synthetic polymer product may exist, for purposes of a specific use. Changing the conditions of polymerization alters the chemical composition of the product and its properties. These alterations include the chain length, or branching, or the tacticity.

With a single monomer as a start, the product is a homopolymer.

Secondary component(s) may be added to create a heteropolymer (co-polymer) and the degree of clustering of the different components can also be controlled.

Physical characteristics, such as hardness, density, mechanical or tensile strength, abrasion resistance, heat resistance, transparency, colour, etc. will depend on the final composition.

Biomolecules

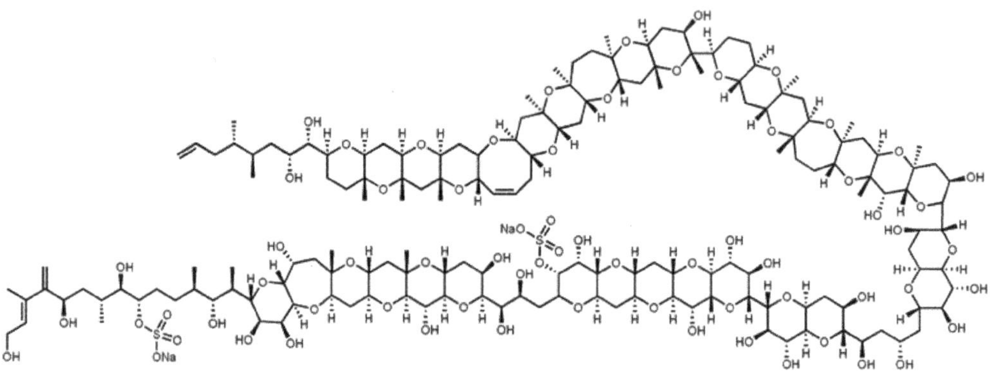

Maitotoxin, a complex organic biological toxin.

Biomolecular chemistry is a major category within organic chemistry which is frequently studied by biochemists. Many complex multi-functional group molecules are important in living organisms. Some are long-chain biopolymers, and these include peptides, DNA, RNA and the polysaccharides such as starches in animals and celluloses in plants.

The other main classes are amino acids (monomer building blocks of peptides and proteins), carbohydrates (which includes the polysaccharides), the nucleic acids (which include DNA and RNA as polymers), and the lipids. In addition, animal biochemistry contains many small molecule intermediates which assist in energy production through the Krebs cycle, and produces isoprene, the most common hydrocarbon in animals. Isoprenes in animals form the important steroid structural (cholesterol) and steroid hormone compounds; and in plants form terpenes, terpenoids, some alkaloids, and a class of hydrocarbons called biopolymer polyisoprenoids present in the latex of various species of plants, which is the basis for making rubber.

Small Molecules

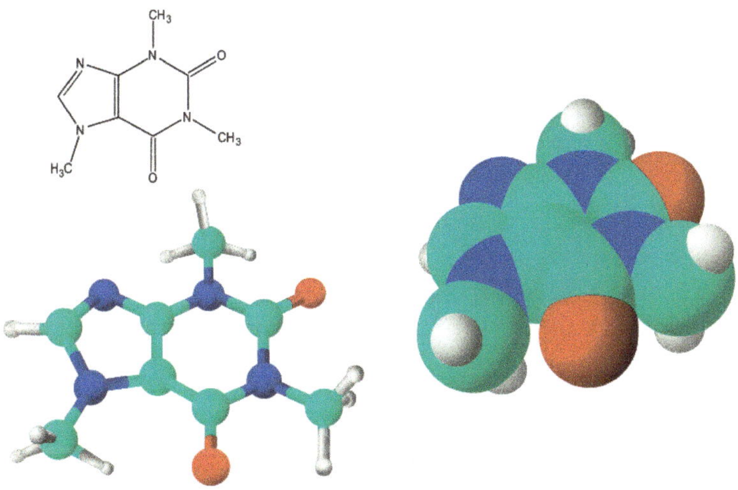

Molecular models of caffeine.

In pharmacology, an important group of organic compounds is small molecules, also referred to as 'small organic compounds'. In this context, a small molecule is a small organic compound that is biologically active, but is not a polymer. In practice, small molecules have a molar mass less than approximately 1000 g/mol.

Fullerenes

Fullerenes and carbon nanotubes, carbon compounds with spheroidal and tubular structures, have stimulated much research into the related field of materials science. The first fullerene was discovered in 1985 by Sir Harold W. Kroto of the United Kingdom and by Richard E. Smalley and Robert F. Curl, Jr., of the United States. Using a laser to vaporize graphite rods in an atmosphere of helium gas, these chemists and their assistants obtained cagelike molecules composed of 60 carbon atoms (C60) joined together by single and double bonds to form a hollow sphere with 12 pentagonal and 20 hexagonal faces—a design that resembles a football, or soccer ball. In 1996 the trio was awarded the Nobel

Prize for their pioneering efforts. The C60 molecule was named buckminsterfullerene (or, more simply, the buckyball) after the American architect R. Buckminster Fuller, whose geodesic dome is constructed on the same structural principles.

Others

Organic compounds containing bonds of carbon to nitrogen, oxygen and the halogens are not normally grouped separately. Others are sometimes put into major groups within organic chemistry and discussed under titles such as organosulfur chemistry, organometallic chemistry, organophosphorus chemistry and organosilicon chemistry.

Organic Reactions

Organic reactions are chemical reactions involving organic compounds. Many of these reactions are associated with functional groups. The general theory of these reactions involves careful analysis of such properties as the electron affinity of key atoms, bond strengths and steric hindrance. These factors can determine the relative stability of short-lived reactive intermediates, which usually directly determine the path of the reaction.

The basic reaction types are: addition reactions, elimination reactions, substitution reactions, pericyclic reactions, rearrangement reactions and redox reactions. An example of a common reaction is a substitution reaction written as:

$$Nu^{-} + C\text{-}X \rightarrow C\text{-}Nu + X^{-}$$

where X is some functional group and Nu is a nucleophile.

The number of possible organic reactions is basically infinite. However, certain general patterns are observed that can be used to describe many common or useful reactions. Each reaction has a stepwise reaction mechanism that explains how it happens in sequence—although the detailed description of steps is not always clear from a list of reactants alone.

The stepwise course of any given reaction mechanism can be represented using arrow pushing techniques in which curved arrows are used to track the movement of electrons as starting materials transition through intermediates to final products.

Organic Synthesis

Synthetic organic chemistry is an applied science as it borders engineering, the "design, analysis, and/or construction of works for practical purposes". Organic synthesis of a novel compound is a problem solving task, where a synthesis is designed for a target molecule by selecting optimal reactions from optimal starting materials. Complex compounds can have tens of reaction steps that sequentially build the desired molecule. The synthesis proceeds by utilizing the reactivity of the functional groups in the molecule. For example, a carbonyl compound can be used as a nucleophile by converting it into an enolate, or as an electrophile; the combination of the two is called the aldol reaction. Designing practically useful syntheses always requires conducting the actual synthesis in the laboratory. The scientific practice of creating novel synthetic routes for complex molecules is called total synthesis.

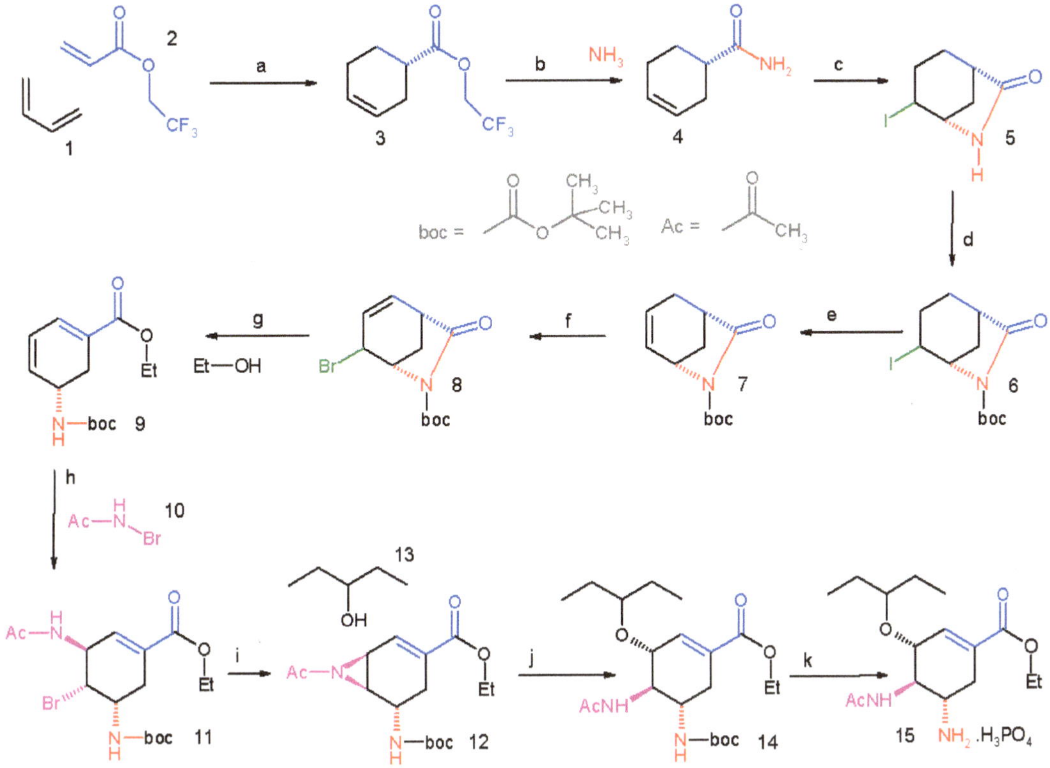

A synthesis designed by E.J. Corey for oseltamivir (Tamiflu). This synthesis has 11 distinct reactions.

Strategies to design a synthesis include retrosynthesis, popularized by E.J. Corey, starts with the target molecule and splices it to pieces according to known reactions. The pieces, or the proposed precursors, receive the same treatment, until available and ideally inexpensive starting materials are reached. Then, the retrosynthesis is written in the opposite direction to give the synthesis. A "synthetic tree" can be constructed, because each compound and also each precursor has multiple syntheses.

REFERENCES

1. Clayden, J.; Greeves, N. and Warren, S. (2012) Organic Chemistry. Oxford University Press. pp. 1–15. ISBN 0199270295.
2. Elschenbroich, C. (2006) Organometallics 3rd Ed., Wiley-VCH
3. Morrison, Robert T.; Boyd, Robert N. and Boyd, Robert K. (1992) Organic Chemistry, 6th ed., Benjamin Cummings.
4. Greenwood, Norman N.; Earnshaw, Alan (1997). Chemistry of the Elements (2nd ed.). Butterworth-Heinemann. ISBN 0-08-037941-9.

5. Henry Marshall Leicester; Herbert S. Klickstein (1951). A Source Book in Chemistry, 1400-1900. Harvard University Press. p. 309.
6. Kiefer, D. M. (1993). "Organic Chemicals' Mauve Beginning". Chem. Eng. News. **71** *(32): 22–23. doi:10.1021/cen-v071n032.p022.*
7. «August Kekulé and Archibald Scott Couper". Chemical Heritage Foundation. Retrieved 27 January 2015.
8. Roberts, Laura (7 December 2010) History of Aspirin. The Telegraph
9. Bosch F & Rosich L (2008). "The contributions of Paul Ehrlich to pharmacology: A tribute on the occasion of the centenary of his Nobel Prize". Pharmacology. **82** *(3): 171–9.* doi:10.1159/000149583. PMC 2790789. PMID 18679046.
10. «Paul Ehrlich, the Rockefeller Institute, and the first targeted chemotherapy". Rockefeller University. Retrieved 3 Aug 2012.
11. «Paul Ehrlich". Chemical Heritage Foundation. Retrieved 29 November 2016.
12. Torker, Sebastian; MüLler, Andre; Sigrist, Raphael; Chen, Peter (2010). "Tuning the Steric Properties of a Metathesis Catalyst for Copolymerization of Norbornene and Cyclooctene toward Complete Alternation". Organometallics. **29** *(12): 2735–2751.* doi:10.1021/om100185g.
13. Steingruber, Elmar (2004) "Indigo and Indigo Colorants" in Ullmann's Encyclopedia of Industrial Chemistry, Wiley-VCH, Weinheim. doi: 10.1002/14356007.a14_149.pub2
14. Nicolaou, K. C.; Sorensen, E. J. (1996). Classics in Total Synthesis: Targets, Strategies, Methods. Wiley. ISBN 978-3-527-29231-8.
15. Allan, Barbara. Livesey, Brian (1994). How to Use Biological Abstracts, Chemical Abstracts and Index Chemicus. Gower. ISBN 978-0566075568
16. Shriner, R.L.; Hermann, C.K.F.; Morrill, T.C.; Curtin, D.Y. and Fuson, R.C. (1997) The Systematic Identification of Organic Compounds. John Wiley & Sons, ISBN 0-471-59748-1
17. "industrial polymers, chemistry of." Encyclopædia Britannica. 2006

CHAPTER 7

Agricultural Chemicals

Chapter Contents

- **Agrochemical**
- **Ecology**
- **Companies**
- **Insecticide**
- **Pesticide**

The Agricultural Chemicals Regulation Law defines "agricultural chemicals" as chemical agents such as fungicides and insecticides that are used to control crop-harming organisms (e.g., fungi, nematodes, mites, insects, and rodents) or viruses (hereinafter collectively referred to as "diseases and pests") (the "crop," as used herein, shall include wood and agroforestry products and those used to promote or inhibit the physiology of agricultural and other products, such as plant growth regulators and germination inhibitors (Hereinafter referred to as "agricultural and other products). The chemical agents here include those manufactured using the above chemical agents as raw materials or ingredients that are intended to control the diseases and pests that are stipulated in the ordinances. The law also includes "natural enemies" and "microorganisms" that are used to control diseases and pests of agricultural and other products, as the agricultural chemicals.

AGROCHEMICAL

Agrochemical or **agrichemical**, a contraction of *agricultural chemical*, is a generic term for the various chemical products used in agriculture. In most cases, *agrichemical* refers to the broad range of pesticides, including insecticides, herbicides, fungicides and nematicides. It may also include synthetic fertilizers, hormones and other chemical growth agents, and concentrated stores of raw animal manure.

The Passaic Agricultural Chemical Works in Newark, New Jersey, 1876.

ECOLOGY

Many agrichemicals are toxic, and agrichemicals in bulk storage may pose significant environmental and/or health risks, particularly in the event of accidental spills. In many countries, use of agrichemicals is highly regulated. Government-issued permits for purchase and use of approved agrichemicals may be required. Significant penalties can result from misuse, including improper storage resulting in spillage. On farms, proper storage facilities and labeling, emergency clean-up equipment and procedures, and safety equipment and procedures for handling, application and disposal are often subject to mandatory standards and regulations. Usually, the regulations are carried out through the registration process.

For instance, bovine somatotropin, though widely used in the U.S.A., is not approved in Canada and some other jurisdictions as there are concerns for the health of cows using it.

COMPANIES

Syngenta was the worldwide leader in agrochemical sales in 2013 at ~$10.9 billion, followed by Bayer CropScience, BASF, Dow Agrosciences, Monsanto, and then DuPont with ~$3.6 billion.

INSECTICIDE

FLIT manual spray pump for insecticides from 1928.

An **insecticide** is a substance used to kill insects. They include ovicides and larvicides used against insect eggs and larvae, respectively. Insecticides are used in agriculture, medicine, industry and by consumers. Insecticides are claimed to be a major factor behind the increase in agricultural 20th century's productivity. Nearly all insecticides have the potential to significantly alter ecosystems; many are toxic to humans; some concentrate along the food chain.

Insecticides can be classified in two major groups: systemic insecticides, which have residual or long term activity; and contact insecticides, which have no residual activity.

Furthermore, one can distinguish three types of insecticide. 1. Natural insecticides, such as nicotine, pyrethrum and neem extracts, made by plants as defenses against insects. 2. Inorganic insecticides, which are metals. 3. Organic insecticides, which are organic chemical compounds, mostly working by contact.

The mode of action describes how the pesticide kills or inactivates a pest. It provides another way of classifying insecticides. Mode of action is important in understanding whether an insecticide will be toxic to unrelated species, such as fish, birds and mammals.

Insecticides are distinct from insect repellents, which do not kill.

Type of Activity

Systemic insecticides become incorporated and distributed systemically throughout the whole plant. When insects feed on the plant, they ingest the insecticide. Systemic insecticides produced by transgenic plants are called plant-incorporated protectants (PIPs). For instance, a gene that codes for a specific Bacillus thuringiensis biocidal protein was introduced into corn and other species. The plant manufactures the protein, which kills the insect when consumed.

Contact insecticides are toxic to insects upon direct contact. These can be inorganic insecticides, which are metals and include arsenates, copper and fluorine compounds, which are less commonly used, and the commonly used sulfur. Contact insecticides can be organic insecticides, i.e. organic chemical compounds, synthetically produced, and comprising the largest numbers of pesticides used today. Or they can be natural compounds like pyrethrum, neem oil etc. Contact insecticides usually have no residual activity.

Efficacy can be related to the quality of pesticide application, with small droplets, such as aerosols often improving performance.

Biological Pesticides

Many organic compounds are produced by plants for the purpose of defending the host plant from predation. A trivial case is tree rosin, which is a natural insecticide. Specific, the production of oleoresin by conifer species is a component of the defense response against insect attack and fungal pathogen infection. Many fragrances, e.g. oil of wintergreen, are in fact antifeedants.

Four extracts of plants are in commercial use: pyrethrum, rotenone, neem oil, and various essential oils

Other Biological Approaches

Plant-Incorporated Protectants

Transgenic crops that act as insecticides began in 1996 with a genetically modified potato that produced the Cry protein, derived from the bacterium Bacillus thuringiensis, which is toxic to beetle larvae such as the Colorado potato beetle. The technique has been expanded to include the use of RNA interference RNAi that fatally silences crucial insect genes. RNAi likely evolved as a defense against viruses. Midgut cells in many larvae take up the molecules and help spread the signal. The technology can target only insects that have the silenced sequence, as was demonstrated when a particular RNAi affected only one of four fruit fly species. The technique is expected to replace many other insecticides, which are losing effectiveness due to the spread of pesticide resistance.

Enzymes

Many plants exude substances to repel insects. Premier examples are substances activated by the enzyme myrosinase. This enzyme converts glucosinolates to various compounds that are toxic to herbivorous insects. One product of this enzyme is allyl isothiocyanate, the pungent ingredient in horseradish sauces.

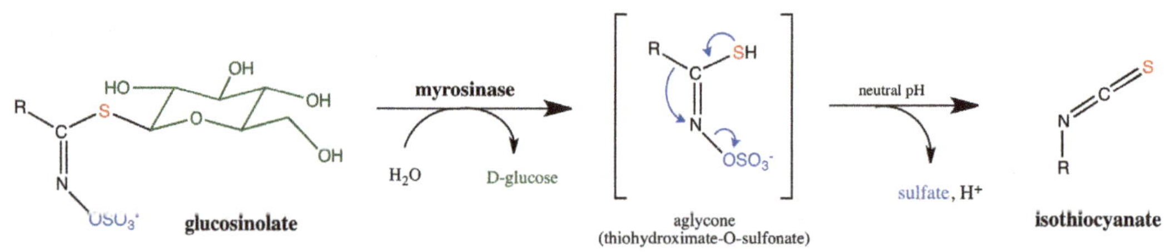

Biosynthesis of antifeedants by the action of myrosinase.

The myrosinase is released only upon crushing the flesh of horseradish. Since allyl isothiocyanate is harmful to the plant as well as the insect, it is stored in the harmless form of the glucosinolate, separate from the myrosinase enzyme.

Bacterial

Bacillus thuringiensis is a bacterial disease that affects Lepidopterans and some other insects. Toxins produced by strains of this bacterium are used as a larvicide against caterpillars, beetles, and mosquitoes. Toxins from *Saccharopolyspora spinosa* are isolated from fermentations and sold as Spinosad. Because these toxins have little effect on other organisms, they are considered more environmentally friendly than synthetic pesticides. The toxin from *B. thuringiensis* (Bt toxin) has been incorporated directly into plants through the use of genetic engineering. Other biological insecticides include products based on entomopathogenic fungi (e.g., *Beauveria bassiana*, *Metarhizium anisopliae*), nematodes (e.g., *Steinernema feltiae*) and viruses (e.g., *Cydia pomonella* granulovirus).

Synthetic Insecticide

A major emphasis of organic chemistry is the development of chemical tools to enhance agricultural productivity. Insecticides represent a major area of emphasis. Many of the major insecticides are inspired by biological analogues. Many others are completely alien to nature.

Organochlorides

The best known organochloride, DDT, was created by Swiss scientist Paul Müller. For this discovery, he was awarded the 1948 Nobel Prize for Physiology or Medicine. DDT was introduced in 1944. It functions by opening sodium channels in the insect's nerve cells. The contemporaneous rise of the chemical industry facilitated large-scale production of DDT and related chlorinated hydrocarbons.

Organophosphates and Carbamates

Organophosphates are another large class of contact insecticides. These also target the insect's nervous system. Organophosphates interfere with the enzymes acetylcholinesterase and other cholinesterases, disrupting nerve impulses and killing or disabling the insect. Organophosphate insecticides and chemical warfare nerve agents (such as sarin, tabun, soman, and VX) work in the same way. Organophosphates have a cumulative toxic effect to wildlife, so multiple exposures to the chemicals amplifies the toxicity. In the US, organophosphate use declined with the rise of substitutes.

Carbamate insecticides have similar mechanisms to organophosphates, but have a much shorter duration of action and are somewhat less toxic.

Pyrethroids

Pyrethroid pesticides mimic the insecticidal activity of the natural compound pyrethrum, the biopesticide found in pyrethrins. These compounds are nonpersistent sodium channel modulators and are less toxic than organophosphates and carbamates. Compounds in this group are often applied against household pests.

Neonicotinoids

Neonicotinoids are synthetic analogues of the natural insecticide nicotine (with much lower acute mammalian toxicity and greater field persistence). These chemicals are acetylcholine receptor agonists. They are broad-spectrum systemic insecticides, with rapid action (minutes-hours). They are applied as sprays, drenches, seed and soil treatments. Treated insects exhibit leg tremors, rapid wing motion, stylet withdrawal (aphids), disoriented movement, paralysis and death. Imidacloprid may be the most common. It has recently come under scrutiny for allegedly pernicious effects on honeybees and its potential to increase the susceptibility of rice to planthopper attacks.

Ryanoids

Ryanoids are synthetic analogues with the same mode of action as ryanodine, a naturally occurring insecticide extracted from *Ryania speciosa* (Flacourtiaceae). They bind to calcium channels in cardiac and skeletal muscle, blocking nerve transmission. Only one such insecticide is currently registered, Rynaxypyr, generic name chlorantraniliprole.

Insect Growth Regulators

Insect growth regulator (IGR) is a term coined to include insect hormone mimics and an earlier class of chemicals, the benzoylphenyl ureas, which inhibit chitin(exoskeleton) biosynthesis in insects. Diflubenzuron is a member of the latter class, used primarily to control caterpillars that are pests. The most successful insecticides in this class are the juvenoids (juvenile hormone analogues). Of these, methoprene is most widely used. It has no observable acute toxicity in rats and is approved by World Health Organization (WHO) for use in drinking water cisterns to combat malaria. Most of its uses are to combat insects where the adult is the pest, including mosquitoes, several fly species, and fleas. Two very similar products, hydroprene and kinoprene, are used for controlling species such as cockroaches and white flies. Methoprene was registered with the EPA in 1975. Virtually no reports of resistance have been filed. A more recent type of IGR is the ecdysone agonist tebufenozide (MIMIC), which is used in forestry and other applications for control of caterpillars, which are far more sensitive to its hormonal effects than other insect orders.

Environmental Effects

Effects on Nontarget Species

Some insecticides kill or harm other creatures in addition to those they are intended to kill. For example, birds may be poisoned when they eat food that was recently sprayed with insecticides or when they mistake an insecticide granule on the ground for food and eat it.

Sprayed insecticide may drift from the area to which it is applied and into wildlife areas, especially when it is sprayed aerially.

DDT

The development of DDT was motivated by desire to replace more dangerous or less effective alternatives. DDT was introduced to replace lead and arsenic-based compounds, which were in widespread use in the early 1940s.

DDT was brought to public attention by Rachel Carson's book *Silent Spring*. One side-effect of DDT is to reduce the thickness of shells on the eggs of predatory birds. The shells sometimes become too thin to be viable, reducing bird populations. This occurs with DDT and related compounds due to the process of bioaccumulation, wherein the chemical, due to its stability and fat solubility, accumulates in organisms' fatty tissues. Also, DDT may biomagnify, which causes progressively higher concentrations in the body fat of animals farther up the food chain. The near-worldwide ban on agricultural use of DDT and related chemicals has allowed some of these birds, such as the peregrine falcon, to recover in recent years. A number of organochlorine pesticides have been banned from most uses worldwide. Globally they are controlled via the Stockholm Convention on persistent organic pollutants. These include: aldrin, chlordane, DDT, dieldrin, endrin, heptachlor, mirex and toxaphene.

Pollinator Decline

Insecticides can kill bees and may be a cause of pollinator decline, the loss of bees that pollinate plants, and colony collapse disorder (CCD), in which worker bees from a beehive or Western honey bee colony abruptly disappear. Loss of pollinators means a reduction in crop yields.Sublethal doses of insecticides (i.e. imidacloprid and other neonicotinoids) affect bee foraging behavior. However, research into the causes of CCD was inconclusive as of June 2007.

PESTICIDE

A crop-duster spraying pesticide on a field.

A Lite-Trac four-wheeled self-propelled crop sprayer spraying pesticide on a field.

Pesticides are substances meant for attracting, seducing, and then destroying any pest. They are a class of biocide. The most common use of pesticides is as plant protection products (also known as crop protection products), which in general protect plants from damaging influences such as weeds, fungi, or insects. This use of pesticides is so common that the term *pesticide* is often treated as synonymous with *plant protection product*, although it is in fact a broader term, as pesticides are also used for non-agricultural purposes. The term pesticide includes all of the following: herbicide, insecticide, insect growth regulator, nematicide, termiticide, molluscicide, piscicide, avicide, rodenticide, predacide, bactericide, insect repellent, animal repellent, antimicrobial, fungicide, disinfectant (antimicrobial), and sanitizer.

In general, a pesticide is a chemical or biological agent (such as a virus, bacterium, antimicrobial, or disinfectant) that deters, incapacitates, kills, or otherwise discourages pests. Target pests can include insects, plant pathogens, weeds, mollusks, birds, mammals, fish, nematodes (roundworms), and microbes that destroy property, cause nuisance, or spread disease, or are disease vectors. Although pesticides have benefits, some also have drawbacks, such as potential toxicity to humans and other species. According to the Stockholm Convention on Persistent Organic Pollutants, 9 of the 12 most dangerous and persistent organic chemicals are organochlorine pesticides.

Type of pesticide	Target pest group
Herbicides	Plant
Algicides or Algaecides	Algae
Avicides	Birds
Bactericides	Bacteria
Fungicides	Fungi and Oomycetes
Insecticides	Insects
Miticides or Acaricides	Mites
Molluscicides	Snails
Nematicides	Nematodes
Rodenticides	Rodents
Virucides	Viruses

The Food and Agriculture Organization (FAO) has defined *pesticide* as: any substance or mixture of substances intended for preventing, destroying, or controlling any pest, including vectors of human or animal disease, unwanted species of plants or animals, causing harm during or otherwise interfering with the production, processing, storage, transport, or marketing of food, agricultural commodities, wood and wood products or animal feedstuffs, or substances that may be administered to animals for the control of insects, arachnids, or other pests in or on their bodies. The term includes substances intended for use as a plant growth regulator, defoliant, desiccant, or agent for thinning fruit or preventing the premature fall of fruit. Also used as substances applied to crops either before or after harvest to protect the commodity from deterioration during storage and transport.

Pesticides can be classified by target organism (e.g., herbicides, insecticides, fungicides, rodenticides, and pediculicides - see table), chemical structure (e.g., organic, inorganic, synthetic, or biological (biopesticide), although the distinction can sometimes blur), and physical state (e.g. gaseous (fumigant)). Biopesticides include microbial pesticides and biochemical pesticides. Plant-derived pesticides, or “botanicals”, have been developing quickly. These include the pyrethroids, rotenoids, nicotinoids, and a fourth group that includes strychnine and scilliroside.[:15]

Many pesticides can be grouped into chemical families. Prominent insecticide families include organochlorines, organophosphates, and carbamates. Organochlorine hydrocarbons (e.g., DDT) could be separated into dichlorodiphenylethanes, cyclodiene compounds, and other related compounds. They operate by disrupting the sodium/potassium balance of the nerve fiber, forcing the nerve to transmit continuously. Their toxicities vary greatly, but they have been phased out because of their persistence and potential to bioaccumulate.[:239–240] Organophosphate and carbamates largely replaced organochlorines. Both operate through inhibiting the enzyme acetylcholinesterase, allowing acetylcholine to transfer nerve impulses indefinitely and causing a variety of symptoms such as weakness or paralysis. Organophosphates are quite toxic to vertebrates, and have in some cases been replaced by less toxic carbamates.[:136–137] Thiocarbamate and dithiocarbamates are subclasses of carbamates. Prominent families of herbicides include phenoxy and benzoic acid herbicides (e.g. 2,4-D), triazines (e.g., atrazine), ureas (e.g., diuron), and Chloroacetanilides (e.g., alachlor). Phenoxy compounds tend to selectively kill broad-leaf weeds rather than grasses. The phenoxy and benzoic acid herbicides function similar to plant growth hormones, and grow cells without normal cell division, crushing the plant's nutrient transport system.[:300] Triazines interfere with photosynthesis.[:335] Many commonly used pesticides are not included in these families, including glyphosate.

Pesticides can be classified based upon their biological mechanism function or application method. Most pesticides work by poisoning pests. A systemic pesticide moves inside a plant following absorption by the plant. With insecticides and most fungicides, this movement is usually upward (through the xylem) and outward. Increased efficiency may be a result. Systemic insecticides, which poison pollen and nectar in the flowers, may kill bees and other needed pollinators.

In 2009, the development of a new class of fungicides called paldoxins was announced. These work by taking advantage of natural defense chemicals released by plants called phytoalexins, which fungi then detoxify using enzymes. The paldoxins inhibit the fungi's detoxification enzymes. They are believed to be safer and greener.

Uses

Pesticides are used to control organisms that are considered to be harmful. For example, they are used to kill mosquitoes that can transmit potentially deadly diseases like West Nile virus, yellow fever, and malaria. They can also kill bees, wasps or ants that can cause allergic reactions. Insecticides can protect animals from illnesses that can be caused by parasites such as fleas. Pesticides can prevent sickness in humans that could be caused by moldy food or diseased produce. Herbicides can be used to clear roadside weeds, trees and brush. They can also kill invasive weeds that may cause environmental damage. Herbicides are commonly applied in ponds and lakes to control algae and plants such as water grasses that can interfere with activities like swimming and fishing and cause the water to look or smell unpleasant. Uncontrolled pests such as termites and mold can damage structures such as houses. Pesticides are used in grocery stores and food storage facilities to manage rodents and insects that infest food such as grain. Each use of a pesticide carries some associated risk. Proper pesticide use decreases these associated risks to a level deemed acceptable by pesticide regulatory agencies such as the United States Environmental Protection Agency (EPA) and the Pest Management Regulatory Agency (PMRA) of Canada.

DDT, sprayed on the walls of houses, is an organochlorine that has been used to fight malaria since the 1950s. Recent policy statements by the World Health Organization have given stronger support to this approach. However, DDT and other organochlorine pesticides have been banned in most countries worldwide because of their persistence in the environment and human toxicity. DDT use is not always effective, as resistance to DDT was identified in Africa as early as 1955, and by 1972 nineteen species of mosquito worldwide were resistant to DDT.

Amount Used

In 2006 and 2007, the world used approximately 2.4 megatonnes (5.3×10^9 lb) of pesticides, with herbicides constituting the biggest part of the world pesticide use at 40%, followed by insecticides (17%) and fungicides (10%). In 2006 and 2007 the U.S. used approximately 0.5 megatonnes (1.1×10^9 lb) of pesticides, accounting for 22% of the world total, including 857 million pounds (389 kt) of conventional pesticides, which are used in the agricultural sector (80% of conventional pesticide use) as well as the industrial, commercial, governmental and home & garden sectors.Pesticides are also found in majority of U.S. households with 78 million out of the 105.5 million households indicating that they use some form of pesticide. As of 2007, there were more than 1,055 active ingredients registered as pesticides, which yield over 20,000 pesticide products that are marketed in the United States.

The US used some 1 kg (2.2 pounds) per hectare of arable land compared with: 4.7 kg in China, 1.3 kg in the UK, 0.1 kg in Cameroon, 5.9 kg in Japan and 2.5 kg in Italy. Insecticide use in the US has declined by more than half since 1980, (.6%/yr) mostly due to the near phase-out of organophosphates. In corn fields, the decline was even steeper, due to the switchover to transgenic Bt corn.

For the global market of crop protection products, market analysts forecast revenues of over 52 billion US$ in 2019.

Benefits

Pesticides can save farmers' money by preventing crop losses to insects and other pests; in the U.S., farmers get an estimated fourfold return on money they spend on pesticides. One study found that not using pesticides reduced crop yields by about 10%. Another study, conducted in 1999, found that a ban on pesticides in the United States may result in a rise of food prices, loss of jobs, and an increase in world hunger.

There are two levels of benefits for pesticide use, primary and secondary. Primary benefits are direct gains from the use of pesticides and secondary benefits are effects that are more long-term.

Primary Benefits

- Controlling pests and plant disease vectors
    - Improved crop/livestock yields
    - Improved crop/livestock quality
    - Invasive species controlled

- Controlling human/livestock disease vectors and nuisance organisms
  - Human lives saved and suffering reduced
  - Animal lives saved and suffering reduced
  - Diseases contained geographically
- Controlling organisms that harm other human activities and structures
  - Drivers view unobstructed
  - Tree/brush/leaf hazards prevented
  - Wooden structures protected

Monetary

Every dollar ($1) that is spent on pesticides for crops yields four dollars ($4) in crops saved. This means based that, on the amount of money spent per year on pesticides, $10 billion, there is an additional $40 billion savings in crop that would be lost due to damage by insects and weeds. In general, farmers benefit from having an increase in crop yield and from being able to grow a variety of crops throughout the year. Consumers of agricultural products also benefit from being able to afford the vast quantities of produce available year-round. The general public also benefits from the use of pesticides for the control of insect-borne diseases and illnesses, such as malaria. The use of pesticides creates a large job market within the agrichemical sector.

Costs

On the cost side of pesticide use there can be costs to the environment, costs to human health, as well as costs of the development and research of new pesticides.

Health Effects

A sign warning about potential pesticide exposure.

Pesticides may cause acute and delayed health effects in people who are exposed. Pesticide exposure can cause a variety of adverse health effects, ranging from simple irritation of the skin and eyes to more severe effects such as affecting the nervous system, mimicking hormones causing reproductive problems, and also causing cancer. A 2007 systematic review found that "most studies on non-Hodgkin lymphoma and leukemia showed positive associations with pesticide exposure" and thus concluded that cosmetic use of pesticides should be decreased. There is substantial evidence of associations between organophosphate insecticide exposures and neurobehavioral alterations. Limited evidence also exists for other negative outcomes from pesticide exposure including neurological, birth defects, and fetal death.

The American Academy of Pediatrics recommends limiting exposure of children to pesticides and using safer alternatives:

The World Health Organization and the UN Environment Programme estimate that each year, 3 million workers in agriculture in the developing world experience severe poisoning from pesticides, about 18,000 of whom die. Owing to inadequate regulation and safety precautions, 99% of pesticide related deaths occur in developing countries that account for only 25% of pesticide usage.According to one study, as many as 25 million workers in developing countries may suffer mild pesticide poisoning yearly. There are several careers aside from agriculture that may also put individuals at risk of health effects from pesticide exposure including pet groomers, groundskeepers, and fumigators.

One study found pesticide self-poisoning the method of choice in one third of suicides worldwide, and recommended, among other things, more restrictions on the types of pesticides that are most harmful to humans.

A 2014 epidemiological review found associations between autism and exposure to certain pesticides, but noted that the available evidence was insufficient to conclude that the relationship was causal.

Environmental Effect

Pesticide use raises a number of environmental concerns. Over 98% of sprayed insecticides and 95% of herbicides reach a destination other than their target species, including non-target species, air, water and soil. Pesticide drift occurs when pesticides suspended in the air as particles are carried by wind to other areas, potentially contaminating them. Pesticides are one of the causes of water pollution, and some pesticides are persistent organic pollutants and contribute to soil contamination.

In addition, pesticide use reduces biodiversity, contributes to pollinator decline, destroys habitat (especially for birds), and threatens endangered species. Pests can develop a resistance to the pesticide (pesticide resistance), necessitating a new pesticide. Alternatively a greater dose of the pesticide can be used to counteract the resistance, although this will cause a worsening of the ambient pollution problem.

Since chlorinated hydrocarbon pesticides dissolve in fats and are not excreted, organisms tend to retain them almost indefinitely. Biological magnification is the process whereby these chlorinated hydrocarbons (pesticides) are more concentrated at each level of the food chain. Among marine animals, pesticide concentrations are higher in carnivorous fishes, and even more so in the fish-eating birds and mammals at the top of the ecological pyramid. Global distillation is the process whereby pesticides are transported from warmer to colder regions of the Earth, in particular the Poles and mountain tops. Pesticides that evaporate into the atmosphere at relatively high temperature can be carried considerable distances (thousands of kilometers) by the wind to an area of lower temperature, where they condense and are carried back to the ground in rain or snow.

In order to reduce negative impacts, it is desirable that pesticides be degradable or at least quickly deactivated in the environment. Such loss of activity or toxicity of pesticides is due to both innate chemical properties of the compounds and environmental processes or conditions. For example, the presence of halogens within a chemical structure often slows down degradation in an aerobic environment. Adsorption to soil may retard pesticide movement, but also may reduce bioavailability to microbial degraders.

Economics

Harm	Annual US cost
Public health	$1.1 billion
Pesticide resistance in pest	$1.5 billion
Crop losses caused by pesticides	$1.4 billion
Bird losses due to pesticides	$2.2 billion
Groundwater contamination	$2.0 billion
Other costs	$1.4 billion
Total costs	**$9.6 billion**

Human health and environmental cost from pesticides in the United States is estimated at $9.6 billion offset by about $40 billion in increased agricultural production:

Additional costs include the registration process and the cost of purchasing pesticides. The registration process can take several years to complete (there are 70 different types of field test) and can cost $50–70 million for a single pesticide. Annually the United States spends $10 billion on pesticides.

Alternatives

Alternatives to pesticides are available and include methods of cultivation, use of biological pest controls (such as pheromones and microbial pesticides), genetic engineering, and methods of interfering with insect breeding. Application of composted yard waste has also been used as a way of controlling pests. These methods are becoming increasingly popular and often are safer than traditional chemical pesticides. In addition, EPA is registering reduced-risk conventional pesticides in increasing numbers.

Cultivation practices include polyculture (growing multiple types of plants), crop rotation, planting crops in areas where the pests that damage them do not live, timing planting according to when pests will be least problematic, and use of trap crops that attract pests away from the real crop.Trap crops have successfully controlled pests in some commercial agricultural systems while reducing pesticide usage; however, in many other systems, trap crops can fail to reduce pest densities at a commercial scale, even when the trap crop works in controlled experiments. In the U.S., farmers have had success controlling insects by spraying with hot water at a cost that is about the same as pesticide spraying.

Release of other organisms that fight the pest is another example of an alternative to pesticide use. These organisms can include natural predators or parasites of the pests. Biological pesticides based on entomopathogenic fungi, bacteria and viruses cause disease in the pest species can also be used.

Interfering with insects' reproduction can be accomplished by sterilizing males of the target species and releasing them, so that they mate with females but do not produce offspring. This technique was first used on the screwworm fly in 1958 and has since been used with the medfly, the tsetse fly, and the gypsy moth. However, this can be a costly, time consuming approach that only works on some types of insects.

Agroecology emphasize nutrient recycling, use of locally available and renewable resources, adaptation to local conditions, utilization of microenvironments, reliance on indigenous knowledge and yield maximization while maintaining soil productivity. Agroecology also emphasizes empowering people and local communities to contribute to development, and encouraging "multi-directional" communications rather than the conventional "top-down" method.

Push Pull Strategy

The term "push-pull" was established in 1987 as an approach for integrated pest management (IPM). This strategy uses a mixture of behavior-modifying stimuli to manipulate the distribution and abundance of insects. "Push" means the insects are repelled or deterred away from whatever resource that is being protected. "Pull" means that certain stimuli (semiochemical stimuli, pheromones, food additives, visual stimuli, genetically altered plants, etc.) are used to attract pests to trap crops where they will be killed. There are numerous different components involved in order to implement a Push-Pull Strategy in IPM.

Many case studies testing the effectiveness of the push-pull approach have been done across the world. The most successful push-pull strategy was developed in Africa for subsistence farming. Another successful case study was performed on the control of *Helicoverpa* in cotton crops in Australia. In Europe, the Middle East, and the United States, push-pull strategies were successfully used in the controlling of *Sitona lineatus* in bean fields.

Some advantages of using the push-pull method are less use of chemical or biological materials and better protection against insect habituation to this control method. Some disadvantages of the push-pull strategy is that if there is a lack of appropriate knowledge of behavioral and chemical ecology of the host-pest interactions then this method becomes unreliable. Furthermore, because the push-pull method is not a very popular method of IPM operational and registration costs are higher.

Effectiveness

Some evidence shows that alternatives to pesticides can be equally effective as the use of chemicals. For example, Sweden has halved its use of pesticides with hardly any reduction in crops. In Indonesia, farmers have reduced pesticide use on rice fields by 65% and experienced a 15% crop increase. A study of Maize fields in northern Florida found that the application of composted yard waste with high carbon to nitrogen ratio to agricultural fields was highly effective at reducing the population of plant-parasitic nematodes and increasing crop yield, with yield increases ranging from 10% to 212%; the observed effects were long-term, often not appearing until the third season of the study.

However, pesticide resistance is increasing. In the 1940s, U.S. farmers lost only 7% of their crops to pests. Since the 1980s, loss has increased to 13%, even though more pesticides are being used. Between 500 and 1,000 insect and weed species have developed pesticide resistance since 1945

Types

Pesticides are often referred to according to the type of pest they control. Pesticides can also be considered as either biodegradable pesticides, which will be broken down by microbes and other living beings into harmless compounds, or persistent pesticides, which may take months or years before they are broken down: it was the persistence of DDT, for example, which led to its accumulation in the food chain and its killing of birds of prey at the top of the food chain. Another way to think about pesticides is to consider those that are chemical pesticides or are derived from a common source or production method. Some examples of chemically-related pesticides are:

Organophosphate Pesticides

Organophosphates affect the nervous system by disrupting, acetylcholinesterase activity, the enzyme that regulates acetylcholine, a neurotransmitter. Most organophosphates are insecticides. They were developed during the early 19th century, but their effects on insects, which are similar to their effects on humans, were discovered in 1932. Some are very poisonous. However, they usually are not persistent in the environment.

Carbamate pesticides

Carbamate pesticides affect the nervous system by disrupting an enzyme that regulates acetylcholine, a neurotransmitter. The enzyme effects are usually reversible. There are several subgroups within the carbamates.

Organochlorine Insecticides

They were commonly used in the past, but many have been removed from the market due to their health and environmental effects and their persistence (e.g., DDT, chlordane, and toxaphene).

Pyrethroid Pesticides

They were Developed as a synthetic version of the naturally occurring pesticide pyrethrin, which is found in chrysanthemums. They have been modified to increase their stability in the environment. Some synthetic pyrethroids are toxic to the nervous system.

Sulfonylurea Herbicides

The following sulfonylureas have been commercialized for weed control: amidosulfuron, azimsulfuron, bensulfuron-methyl, chlorimuron-ethyl, ethoxysulfuron, flazasulfuron, flupyrsulfuron-methyl-sodium, halosulfuron-methyl, imazosulfuron, nicosulfuron, oxasulfuron, primisulfuron-methyl, pyrazosulfuron-ethyl, rimsulfuron, sulfometuron-methyl Sulfosulfuron, terbacil, bispyribac-sodium, cyclosulfamuron, and pyrithiobac-sodium. Nicosulfuron, triflusulfuron methyl, and chlorsulfuron are broad-spectrum herbicides that kill plants by inhibiting the enzyme acetolactate synthase. In the 1960s, more than 1 kg/ha (0.89 lb/acre) crop protection chemical was typically applied, while sulfonylureates allow as little as 1% as much material to achieve the same effect.

Biopesticides

Biopesticides are certain types of pesticides derived from such natural materials as animals, plants, bacteria, and certain minerals. For example, canola oil and baking soda have pesticidal applications and are considered biopesticides. Biopesticides fall into three major classes:

- Microbial pesticides which consist of bacteria, entomopathogenic fungi or viruses (and sometimes includes the metabolites that bacteria or fungi produce). Entomopathogenic nematodes are also often classed as microbial pesticides, even though they are multi-cellular.
- Biochemical pesticides or herbal pesticides are naturally occurring substances that control (or monitor in the case of pheromones) pests and microbial diseases.
- Plant-incorporated protectants (PIPs) have genetic material from other species incorporated into their genetic material (*i.e.* GM crops). Their use is controversial, especially in many European countries.

Classified by Type of Pest

Pesticides that are related to the type of pests are:

Type	Action
Algicides	Control algae in lakes, canals, swimming pools, water tanks, and other sites
Antifouling agents	Kill or repel organisms that attach to underwater surfaces, such as boat bottoms
Antimicrobials	Kill microorganisms (such as bacteria and viruses)

Attractants	Attract pests (for example, to lure an insect or rodent to a trap). (However, food is not considered a pesticide when used as an attractant.)
Biopesticides	Biopesticides are certain types of pesticides derived from such natural materials as animals, plants, bacteria, and certain minerals
Biocides	Kill microorganisms
Disinfectants and sanitizers	Kill or inactivate disease-producing microorganisms on inanimate objects
Fungicides	Kill fungi (including blights, mildews, molds, and rusts)
Fumigants	Produce gas or vapor intended to destroy pests in buildings or soil
Herbicides	Kill weeds and other plants that grow where they are not wanted
Insecticides	Kill insects and other arthropods
Miticides	Kill mites that feed on plants and animals
Microbial pesticides	Microorganisms that kill, inhibit, or out compete pests, including insects or other microorganisms
Molluscicides	Kill snails and slugs
Nematicides	Kill nematodes (microscopic, worm-like organisms that feed on plant roots)
Ovicides	Kill eggs of insects and mites
Pheromones	Biochemicals used to disrupt the mating behavior of insects
Repellents	Repel pests, including insects (such as mosquitoes) and birds
Rodenticides	Control mice and other rodents

Further types of Pesticides

- **Defoliants** : Cause leaves or other foliage to drop from a plant, usually to facilitate harvest.
- **Desiccants** : Promote drying of living tissues, such as unwanted plant tops.
- **Insect growth regulators** : Disrupt the molting, maturity from pupal stage to adult, or other life processes of insects.
- **Plant growth regulators** : Substances (excluding fertilizers or other plant nutrients) that alter the expected growth, flowering, or reproduction rate of plants.

Regulation

International

In most countries pesticides must be approved for sale and use by a government agency.

In Europe, recent EU legislation has been approved banning the use of highly toxic pesticides including those that are carcinogenic, mutagenic or toxic to reproduction, those that are endocrine-disrupting, and those that are persistent, bioaccumulative and toxic (PBT) or very persistent and very bioaccumulative (vPvB). Measures were approved to improve the general safety of pesticides across all EU member states.

Though pesticide regulations differ from country to country, pesticides, and products on which they were used are traded across international borders. To deal with inconsistencies in regulations among countries, delegates to a conference of the United Nations Food and Agriculture Organization adopted an International Code of Conduct on the Distribution and Use of Pesticides in 1985 to create voluntary standards of pesticide regulation for different countries. The Code was updated in 1998 and 2002. The FAO claims that the code has raised awareness about pesticide hazards and decreased the number of countries without restrictions on pesticide use.

Three other efforts to improve regulation of international pesticide trade are the United Nations London Guidelines for the Exchange of Information on Chemicals in International Trade and the United Nations Codex Alimentarius Commission. The former seeks to implement procedures for ensuring that prior informed consent exists between countries buying and selling pesticides, while the latter seeks to create uniform standards for maximum levels of pesticide residues among participating countries. Both initiatives operate on a voluntary basis.

Pesticides safety education and pesticide applicator regulation are designed to protect the public from pesticide misuse, but do not eliminate all misuse. Reducing the use of pesticides and choosing less toxic pesticides may reduce risks placed on society and the environment from pesticide use. Integrated pest management, the use of multiple approaches to control pests, is becoming widespread and has been used with success in countries such as Indonesia, China, Bangladesh, the U.S., Australia, and Mexico. IPM attempts to recognize the more widespread impacts of an action on an ecosystem, so that natural balances are not upset. New pesticides are being developed, including biological and botanical derivatives and alternatives that are thought to reduce health and environmental risks. In addition, applicators are being encouraged to consider alternative controls and adopt methods that reduce the use of chemical pesticides.

Pesticides can be created that are targeted to a specific pest's lifecycle, which can be environmentally more friendly. For example, potato cyst nematodes emerge from their protective cysts in response to a chemical excreted by potatoes; they feed on the potatoes and damage the crop.A similar chemical can be applied to fields early, before the potatoes are planted, causing the nematodes to emerge early and starve in the absence of potatoes.

United States

In the United States, the Environmental Protection Agency (EPA) is responsible for regulating pesticides under the Federal Insecticide, Fungicide, and Rodenticide Act (FIFRA) and the Food Quality Protection Act (FQPA). Studies must be conducted to establish the conditions in which the material is safe to use and the effectiveness against the intended pest(s). The EPA regulates pesticides to ensure that these products do not pose adverse effects to humans or the environment. Pesticides produced before November

1984 continue to be reassessed in order to meet the current scientific and regulatory standards. All registered pesticides are reviewed every 15 years to ensure they meet the proper standards. During the registration process, a label is created. The label contains directions for proper use of the material in addition to safety restrictions. Based on acute toxicity, pesticides are assigned to a Toxicity Class.

Preparation for an application of hazardous herbicide in USA.

Some pesticides are considered too hazardous for sale to the general public and are designated restricted use pesticides. Only certified applicators, who have passed an exam, may purchase or supervise the application of restricted use pesticides.Records of sales and use are required to be maintained and may be audited by government agencies charged with the enforcement of pesticide regulations. These records must be made available to employees and state or territorial environmental regulatory agencies.

The EPA regulates pesticides under two main acts, both of which amended by the Food Quality Protection Act of 1996. In addition to the EPA, the United States Department of Agriculture (USDA) and the United States Food and Drug Administration (FDA) set standards for the level of pesticide residue that is allowed on or in crops. The EPA looks at what the potential human health and environmental effects might be associated with the use of the pesticide.

In addition, the U.S. EPA uses the National Research Council's four-step process for human health risk assessment: (1) Hazard Identification, (2) Dose-Response Assessment, (3) Exposure Assessment, and (4) Risk Characterization.

Recently Kaua'i County (Hawai'i) passed Bill No. 2491 to add an article to Chapter 22 of the county's code relating to pesticides and GMOs. The bill strengthens protections of local communities in Kaua'i where many large pesticide companies test their products.

Since before 2000 BC, humans have utilized pesticides to protect their crops. The first known pesticide was elemental sulfur dusting used in ancient Sumer about 4,500 years ago in ancient Mesopotamia. The Rig Veda, which is about 4,000 years old, mentions the use of poisonous plants for pest control. By the 15th century, toxic chemicals such as arsenic, mercury, and lead were being applied to crops to kill pests. In the 17th century, nicotine sulfate was extracted from tobacco leaves for use as an insecticide. The 19th century saw the introduction of two more natural pesticides, pyrethrum, which is derived from chrysanthemums, and rotenone, which is derived from the roots of tropical vegetables. Until the 1950s, arsenic-based pesticides were dominant. Paul Müller discovered that DDT was a very effective insecticide. Organochlorines such as DDT were dominant, but they were replaced in the U.S. by organophosphates and carbamates by 1975. Since then, pyrethrin compounds have become the dominant insecticide. Herbicides became common in the 1960s, led by "triazine and other nitrogen-based compounds, carboxylic acids such as 2,4-dichlorophenoxyacetic acid, and glyphosate".

The first legislation providing federal authority for regulating pesticides was enacted in 1910; however, decades later during the 1940s manufacturers began to produce large amounts of synthetic pesticides and their use became widespread. Some sources consider the 1940s and 1950s to have been the start of the "pesticide era." Although the U.S. Environmental Protection Agency was established in 1970 and amendments to the pesticide law in 1972, pesticide use has increased 50-fold since 1950 and 2.3 million tonnes (2.5 million short tons) of industrial pesticides are now used each year. Seventy-five percent of all pesticides in the world are used in developed countries, but use in developing countries is increasing. A study of USA pesticide use trends through 1997 was published in 2003 by the National Science Foundation's Center for Integrated Pest Management.

In the 1960s, it was discovered that DDT was preventing many fish-eating birds from reproducing, which was a serious threat to biodiversity. Rachel Carson wrote the best-selling book *Silent Spring* about biological magnification. The agricultural use of DDT is now banned under the Stockholm Convention on Persistent Organic Pollutants, but it is still used in some developing nations to prevent malaria and other tropical diseases by spraying on interior walls to kill or repel mosquitoes.

REFERENCES

1. US Environmental (July 24, 2007), What is a pesticide? epa.gov. Retrieved on September 15, 2007.
2. Carolyn Randall (ed.), et al., National Pesticide Applicator Certification Core Manual (2013) National Association of State Departments of Agriculture Research Foundation, Washington, DC, Ch.1
3. Beginner's guide
4. Gilden RC, Huffling K, Sattler B (January 2010). "Pesticides and health risks". J Obstet Gynecol Neonatal Nurs. 39 *(1): 103–10. doi:10.1111/j.1552-6909.2009.01092.x. PMID 20409108.*
5. Food and Agriculture Organization of the United Nations (2002), International Code of Conduct on the Distribution and Use of Pesticides. Retrieved on 2007-10-25.
6. www.chromatography-online.org".
7. Council on Scientific Affairs, American Medical Association. (1997). Educational

and Informational Strategies to Reduce Pesticide Risks. Preventive Medicine, Volume 26, Number 2

8. EPA. Types of Pesticides. Last updated on Thursday, January 29th, 2009.
9. Kamrin MA. (1997). Pesticide Profiles: toxicity, environmental impact, and fate. CRC Press.
10. Cornell University. Toxicity of pesticides. Pesticide fact sheets and tutorial, module 4. Pesticide Safety Education Program. Retrieved on 2007-10-10.
11. EurekAlert. (2009). New 'green' pesticides are first to exploit plant defenses in battle of the fungi.
12. The benefits of pesticides: A story worth telling. Purdue.edu. Retrieved on September 15, 2007.
13. Helfrich, LA, Weigmann, DL, Hipkins, P, and Stinson, ER (June 1996), Pesticides and aquatic animals: A guide to reducing impacts on aquatic systems. Virginia Cooperative Extension. Retrieved on 2007-10-14.
14. World Health Organization (September 15, 2006), WHO gives indoor use of DDT a clean bill of health for controlling malaria. Retrieved on September 13, 2007.
15. PANNA: PAN Magazine: In Depth: DDT & Malaria
16. A STORY TO BE SHARED: THE SUCCESSFUL FIGHT AGAINST MALARIA IN VIETNAM
17. EPA Pesticide Industry Sales and Usage Report
18. Goldman LR (2007). "Managing pesticide chronic health risks: U.S. Policies". Journal of Agromedicine. 12 *(1): 67–75.* doi:10.1300/J096v12n02_08. PMID 18032337.
19. «CDC - Pesticide Illness & Injury Surveillance - NIOSH Workplace Safety and Health Topic". Cdc.gov. 2013-09-11. Retrieved 2014-01-28.
20. «Infographic: Pesticide Planet". Science. 341 *(6147): 730–731. 2013. doi:10.1126/science.341.6147.730. PMID 23950524.*
21. «Market Study: Crop Protection (UC-2805)". June 2012. Retrieved 2012-08-09.
22. Kellogg RL, Nehring R, Grube A, Goss DW, and Plotkin S (February 2000), Environmental indicators of pesticide leaching and runoff from farm fields. United States Department of Agriculture Natural Resources Conservation Service. Retrieved on 2007-10-03.
23. Kuniuki S (2001). "Effects of organic fertilization and pesticide application on growth and yield of field-grown rice for 10 years". Japanese Journal of Crop Science. 70 *(4): 530–540.*
24. Knutson, R. (1999). Economic Impact of Reduced Pesticide Use in the United States.Agricultural and Food Policy Center. Texas A&M University.
25. Cooper, Jerry and Hans Dobson. "The benefits of pesticides to mankind and the environment" Crop Protection 26 (2007): 1337-1348., Retrieved on February 25, 2011.

26. Pimentel, David, H. Acquay, M. Biltonen, P. Rice, and M. Silva. "Environmental and Economic Costs of Pesticide Use." BioScience 42.10 (1992): 750-60., . Retrieved on February 25, 2011.

27. Fantke P, Friedrich R, Jolliet O (2012). "Health impact and damage cost assessment of pesticides in Europe". Environ Int. 49: 9–17. doi:10.1016/j.envint.2012.08.001. PMID 22940502.

28. U.S. Environmental Protection Agency (August 30, 2007), Pesticides: Health and Safety. National Assessment of the Worker Protection Workshop #3.

29. «Human Health Issues | Pesticides | US EPA". Epa.gov. 2006-06-28. Retrieved 2014-01-28.

30. Bassil KL, Vakil C, Sanborn M, Cole DC, Kaur JS, Kerr KJ (October 2007). "Cancer health effects of pesticides: Systematic review". Can Fam Physician. 53 *(10): 1 u704–11. PMC 2231435*. PMID 17934034.

31. Jurewicz J, Hanke W (2008). "Prenatal and childhood exposure to pesticides and neurobehavioral development: review of epidemiological studies". Int J Occup Med Environ Health. 21 *(2): 121–32.* doi:10.2478/v10001-008-0014-z. PMID 18614459.

32. Weselak M, Arbuckle TE, Foster W (2007). "Pesticide exposures and developmental outcomes: the epidemiological evidence". J Toxicol Environ Health B Crit Rev. 10 *(1-2): 41–80.* doi:10.1080/10937400601034571. PMID 18074304.

33. Wigle DT, Arbuckle TE, Turner MC, et al. (May 2008). "Epidemiologic evidence of relationships between reproductive and child health outcomes and environmental chemical contaminants". J Toxicol Environ Health B Crit Rev. 11 *(5-6): 373–517.* doi:10.1080/10937400801921320. PMID 18470797.

34. Mink PJ, Mandel JS, Lundin JI, Sceurman BK (November 2011). "Epidemiologic studies of glyphosate and non-cancer health outcomes: a review". Regul. Toxicol. Pharmacol. 61 *(2): 172–84. doi:10.1016/j.yrtph.2011.07.006. PMID 21798302.*

35. Sanborn M, Kerr KJ, Sanin LH, Cole DC, Bassil KL, Vakil C (October 2007). "Non-cancer health effects of pesticides: Systematic review and implications for family doctors". Can Fam Physician. 53 *(10): 1712–20. PMC 2231436. PMID 17934035.*

36. «Pesticide exposure in children". Pediatrics. 130 *(6): e1757–63. December 2012. doi:10.1542/peds.2012-2757. PMID 23184103.*

37. Miller GT (2004), Sustaining the Earth, 6th edition. Thompson Learning, Inc. Pacific Grove, California. Chapter 9, Pages 211-216.

38. «www.who.int" (PDF).

39. Jeyaratnam J (1990). "Acute pesticide poisoning: a major global health problem". World Health Stat Q. 43 *(3): 139–44. PMID 2238694.*

40. «CDC - Pesticide Illness & Injury Surveillance - NIOSH Workplace Safety and Health Topic". www.cdc.gov. Retrieved 2016-02-11.

41. Gunnell D, Eddleston M, Phillips MR, Konradsen F (2007). "The global

distribution of fatal pesticide self-poisoning: Systematic review". BMC Public Health. 7 *(1): 357.* doi:10.1186/1471-2458-7-357. PMC 2262093. PMID 18154668.

42. Kalkbrenner AE, Schmidt RJ, Penlesky AC (September 2014). "Environmental Chemical Exposures and Autism Spectrum Disorders: A Review of the Epidemiological Evidence". Curr Probl Pediatr Adolesc Health Care. 44 *(10): 277–318. doi:10.1016/j.cppeds.2014.06.001. PMID 25199954.*
43. Wells, M (March 11, 2007). "Vanishing bees threaten U.S. crops". www.bbc.co.uk. London: BBC News. Retrieved 2007-09-19.
44. Palmer, WE, Bromley, PT, and Brandenburg, RL. Wildlife & pesticides - Peanuts. North Carolina Cooperative Extension Service. Retrieved on 2007-10-11.
45. Castro, Peter, and Michael E.Huber. Marine Biology. 8th. New York: McGraw-Hill Companies Inc., 2010. Print.
46. L. Quinn, Amie. "The impacts of agriculture and temperature on the physiological stress response in fish." Uleth. University of Lethbridge, n.d. Web. 20 Nov 2012.
47. Sims, G. K. and A.M. Cupples. 1999. Factors controlling degradation of pesticides in soil. Pesticide Science 55:598-601.
48. Sims, G. K. and L.E. Sommers. 1986. Biodegradation of pyridine derivatives in soil suspensions. Environmental Toxicology and Chemistry. 5:503-509.
49. Wolt J. D.; Smith J.K.; Sims J.K.; Duebelbeis D.O. (1996). "Products and kinetics of cloramsulam-methyl aerobic soil metabolism". J. Agric. Food Chem. 44: 324–332. doi:10.1021/jf9503570.
50. Pimentel, David. "Environmental and Economic Costs of the Application of Pesticides Primarily in the United States" Environment, Development and Sustainability 7 (2005): 229-252. Retrieved on February 25, 2011.
51. R. McSorley and R. N. Gallaher, "Effect of Yard Waste Compost on Nematode Densities and Maize Yield", J Nematology, Vol. 2, No. 4S, pp. 655–660, Dec. 1996.
52. Shelton, A.m.; Badenes-Perez, F.r. (2005-12-06). "Concepts and applications of trap cropping in pest management". Annual Review of Entomology. 51 *(1): 285–308. doi:10.1146/annurev.ento.51.110104.150959. ISSN 0066-4170.*
53. Holden, Matthew H.; Ellner, Stephen P.; Lee, Doo-Hyung; Nyrop, Jan P.; Sanderson, John P. (2012-06-01). "Designing an effective trap cropping strategy: the effects of attraction, retention and plant spatial distribution". Journal of Applied Ecology. 49 *(3): 715–722. doi:10.1111/j.1365-2664.2012.02137.x. ISSN 1365-2664.*
54. (July 2007), The biological control of pests. Retrieved on September 17, 2007.
55. SP-401 Skylab, Classroom in Space: Part III - Science Demonstrations, Chapter 17: Life Sciences. History.nasa.gov. Retrieved on September 17, 2007.
56. «Principles List". Agroecology.org. Retrieved 2014-05-05.
57. Cook SM, Khan ZR, Pickett JA (2007). "The use of push-pull strategies in integrated pest management". Annu. Rev. Entomol. 52 *(1): 375–400. doi:10.1146/annurev.ento.52.110405.091407. PMID 16968206.*

58. «Pesticides 101 - A primer on pesticides, their use in agriculture and the exposure we face | Pesticide Action Network". Panna.org. Retrieved 2014-01-28.
59. «Types of Pesticides". US Environmental Protection Agency. Retrieved 20 February 2013.
60. Arnold P. Appleby, Franz Müller, Serge Carpy "Weed Control" in Ullmann's Encyclopedia of Industrial Chemistry 2002, Wiley-VCH, Weinheim. doi:10.1002/14356007.a28_165
61. «Nicosulfuron". EXTOXNET. Retrieved 9 May 2013.
62. EFSA September 30, 2008 EFSA Scientific Report (2008) 195, 1-115: Conclusion on the peer review of triflusulfuron
63. Lamberth C, Jeanmart S, Luksch T, Plant A (2013). "Current Challenges and Trends in the Discovery of Agrochemicals". Science. 341 *(6147): 742–6. doi:10.1126/science.1237227. PMID 23950530.*
64. Coombs, Amy. "Fighting Microbes with Microbes". The Scientist. Retrieved 18 April 2013.
65. Francis Borgio J, Sahayaraj K and Alper Susurluk I (eds) . Microbial Insecticides: Principles and Applications, Nova Publishers, USA. 492pp. ISBN 978-1-61209-223-2
66. Pal, GK; Kumar, B (2013). "Antifungal activity of some common weed extracts against wilt causing fungi, Fusarium oxysporum" (PDF). Current Discovery. International Young Scientist Association for Applied Research and Development. 2 *(1): 62–67. ISSN 2320-4400. Retrieved February 8, 2014.*
67. National Pesticide Information Center Last updated November 21, 2013 Plant Incorporated Protectants (PIPs) / Genetically Modified Plants
68. Willson, Harold R (February 23, 1996), Pesticide Regulations. University of Minnesota. Retrieved on 2007-10-15.
69. Pesticide Legislation Approved last retrieved 13 January 2009
70. Food and Agriculture Organization of the United Nations, Programmes: International Code of Conduct on the Distribution and Use of Pesticides. Retrieved on 2007-10-25.
71. Reynolds, JD (1997), International pesticide trade: Is there any hope for the effective regulation of controlled substances? Florida State University Journal of Land Use & Environmental Law, Volume 131. Retrieved on 2007-10-16.
72. Daly H, Doyen JT, and Purcell AH III (1998), Introduction to insect biology and diversity, 2nd edition. Oxford University Press. New York, New York. Chapter 14, Pages 279-300.
73. Science Daily, (October 11, 2001), Environmentally-friendly pesticide to combat potato cyst nematodes. Sciencedaily.com. Retrieved on September 19, 2007.
74. "Pesticides and Public Health | Pesticides | US EPA". Epa.gov. Retrieved 2014-01-28.
75. Data Requirements for Pesticide Registration". US EPA.

76. «Protocol for Conducting Environmental Compliance Audits under the Federal Insecticide, Fungicide, and Rodenticide Act (FIFRA)" (PDF). US EPA.

77. «Restricted-Use Pesticides: Dealer and Applicator Records Inspections" (PDF). US EPA.

78. «Chemical Hazard Communication; U.S. Department of Labor; Occupational Safety and Health Administration". US OSHA.

79. «Emergency Planning and Community Right-To-Know Act (EPCRA)". US EPA.

80. Stephen J. Toth, Jr., "Pesticide Impact Assessment Specialist, North Carolina Cooperative Extension Service, "Federal Pesticide Laws and Regulations" March, 1996. Retrieved on February 25, 2011.

81. US Environmental Protection Agency (February 16, 2011), Pesticide Registration Program epa.gov. Retrieved on February 25, 2011.

82. "Assessing Health Risks from Pesticides". U.S. Environmental Protection Agency]"

83. Bill No. 2491, Draft 2, Council of the County of Kaua'i

84. Rao GV, Rupela OP, Rao VR, Reddy YV (2007). "Role of biopesticides in crop protection: present status and future prospects" (PDF). Indian Journal of Plant Protection. 35 *(1): 1–9.*

85. Miller, GT (2002). Living in the Environment (12th Ed.). Belmont: Wadsworth/ Thomson Learning. ISBN 0-534-37697-5

86. Ritter SR. (2009). Pinpointing Trends In Pesticide Use In 1939. C&E News.

87. Goldman L.R. (2007). "Managing pesticide chronic health risks: U.S. policies". Journal of Agromedicine. 12 *(1): 57–75.* doi:10.1300/J096v12n02_08. PMID 18032337.

88. Graeme Murphy (December 1, 2005), Resistance Management - Pesticide Rotation. Ontario Ministry of Agriculture, Food and Rural Affairs. Retrieved on September 15, 2007.

89. Arnold L. Aspelin (February, 2003), PESTICIDE USAGE IN THE UNITED STATES: Trends During the 20th Century. NSF CIPM Technical Bulletin 105. Retrieved on October 28, 2010.

90. Lobe, J (Sept 16, 2006), "WHO urges DDT for malaria control Strategies," Inter Press Service, cited from Commondreams.org. Retrieved on September 15, 2007.

CHAPTER 8

Fertilizers and Nitrate Leaching

Chapter Contents

Nitrogen is the most limiting nutrient for primary production in terrestrial ecosystems. In crop production, N is added to feed the whole plant, not just the harvestable parts. In permanent crops (trees & vines, asparagus, etc.), the perennial parts use and store significant amounts of nitrogen (roots, corms, trunk, branches, vines) Supplying crops with nitrogen from external sources (i.e. nitrogen fertilization) is essential for sustainable and profitable crop production. Nitrogen fertilizer is an expensive input, making luxury applications very unattractive to growers.

THE NITRATE PROBLEM

Nitrate is one of the facts of life. It is essential for the growth of many plant species, including most of those we eat, but it becomes a problem if it gets into water in which it is not wanted. It is perceived mainly as a chemical fertilizer used by farmers, but much of the nitrate found in soil is produced by the microbes that break down plant residues and other nitrogen-containing residues in the soil. There is no difference between nitrate from fertilizer and that produced by microbes, but, whatever its origin, this rather commonplace chemical entity has now become a major environmental problem and is also treated as a health hazard. Concentrations of nitrate increased in natural waters for two decades before levelling off in the 1980s.1 Applications of nitrogen fertilizer followed a similar pattern,1 so many people drew the obvious conclusion that the increase in nitrate concentrations arose from the greater use of fertilizers. However, conclusions drawn from coincident changes need to be examined carefully. The fact that the birth-rate in Europe declined at the same time as the population of storks does not necessarily mean that storks bring babies! The stork and the birth-rate were probably both responding to the increased size and affluence of the human population and its increased and more intensive use of land for agriculture.Much the same can be said for nitrate concentrations and fertilizer use. It is not so much that more use of fertilizer has led to more nitrate in natural waters, as that increases in both the area of land used for arable agriculture and the intensity with which it is farmed have led to both greater concentrations of nitrate and greater use of nitrogen fertilizer. The fertilizer is part of the intensification package, and examining it as a cause of the nitrate problem without considering the rest of the package could lead to false conclusions.2 This point is particularly important because, as we shall see later, nitrogen fertilizers have both a direct and an indirect role in the nitrate problem. Nitrate is seen as a threat to both public health and natural waters. Of these threats the latter is definitely the more immediate, but the health issue has attracted more public concern.

NITRATE AS A HEALTH HAZARD

Nitrate is not a new problem. Excessive concentrations were recorded in many domestic wells in a survey conducted 100 years ago.3 What is new is the public concern about nitrate. This arises from two medical conditions that have been linked to nitrate: methaemoglobinaemia ('blue-baby syndrome') in infants, and stomach cancer in adults. Both are serious conditions, so we need to examine possible links carefully, but we need to note that these conditions are not caused by nitrate but by the nitrite to which it may be reduced. Nitrate itself is harmless and is most notable from a medical standpoint as a treatment for phosphatic kidney stones. Methaemoglobinaemia. The 'blue-baby syndrome' can occur when an infant less than about one year old ingests too much nitrate. Microbes in the stomach convert the nitrate to nitrite and when this reaches the blood-stream it reacts with the haemoglobin, the molecule that transports oxygen around the body. Normal oxyhaemoglobin, which contains iron in the iron(II) state, becomes methaemoglobin in which the iron is in the iron(III) state, greatly lessening the capacity of the blood to carry oxygen and causing what might be described as chemical suffocation. Very young children are susceptible because foetal haemoglobin, which has a greater affinity for nitrite than normal haemoglobin, persists in the blood-stream for a while, and because their stomachs are not sufficiently acid to inhibit the microbes that convert nitrate to nitrite. Gastroenteritis greatly exacerbates the effects of the nitrite. This condition is usually very rare. In the UK the last case was in 1972 and the last death4 in 1950, but in Hungary there were over 1300 cases between 1976 and 1982.5 One reason for this difference between the two countries may lie in the origin of the water. Practically all known cases were associated with water from wells, and the condition is known as 'well-water methaemoglobinaemia' in the USA. In 98% of these cases the wells were dug privately and may have been too close to disposal points for animal or human excreta, thereby increasing the risk of pollution not only by nitrate but also by E. coli and other organisms that cause gastroenteritis. The author is not aware of any case in which methaemoglobinaemia was caused by tap water from a mains supply such as that used by most households in the UK. That being said, there is no room for complacency about the condition. In the fatal case in 1950, the doctor reported that, 'There were diarrhoea and vomiting and the child's complexion was slate-blue'.4 In a similar but non-fatal case in the same year, 'Blood drawn from a vein was a deep chocolate-brown'.4 Stomach Cancer. Of all the cancers, that of the stomach causes the second largest number of deaths. Only lung cancer kills more men, and only breast cancer kills more women. Stomach cancer is a painful and debilitating way to die, and the link to nitrate in water that has been suggested is a serious matter. There are good theoretical reasons for proposing such a link. Nitrite produced from nitrate could react in the stomach with a secondary amine coming from the breakdown of meat or other protein to produce an N-nitroso compound.

$$R^1R^2N{-}H + NO_2^- + H^+ \longrightarrow R^1R^2N{-}N{=}O + H_2O$$

The N-nitroso compounds are carcinogenic, so the reaction could result in stomach cancer. This mechanism is essentially a hypothetical one, and tests were made to evaluate it. Three tests in particular suggested that the hypothesis was not correct, that is, that there is no clear link between stomach cancer and nitrate in water. One test, made at the Radcliffe Infirmary in Oxford,6 identified two areas of the UK in which the incidence of stomach cancer was particularly high and two in which it was particularly low. People attending hospitals in these areas as visitors rather than patients were asked to provide samples of saliva. The hypothesis suggested that samples from the high-risk areas should contain more nitrite and nitrate than those from the low-risk areas, but this was not so. The samples from the low-risk populations had nitrate concentrations 50% greater than those from the high-risk areas. Another test looked at nitrate concentrations in 229 urban areas in the UK between 1969 and 1973 and at deaths from stomach cancer in the same areas at the same time.7 The hypothesis suggested that there should be a positive relationship between them. The results showed a negative one. The third test is particularly interesting in the context of a paper about nitrogen fertilizer because it involved a plant making it.8 If you work in, or even visit, a plant producing ammonium nitrate or another fertilizer, you can taste fertilizer within minutes of entering, as the fine dust in the air dissolves in your saliva. If anyone is going to get stomach cancer as a result of exposure to nitrate, it is workers in nitrate fertilizer plants, but an epidemiological survey showed that their mortality due to stomach cancer did not differ from that among workers in comparable jobs (Table 1). Twelve deaths occurred from stomach cancer during the 35 years of the study, while 12.06 deaths would have been expected from the mortality rate among the other workers. Although there are a few caveats that need to be noted about these tests,2 the results suggest strongly that there is no real link between stomach cancer and nitrate in water. Further support for this conclusion comes from the fact that, while nitrate concentrations have been increasing in water during the past 30 years, the incidence of stomach cancer has been declining, and about ten years ago the absence of any link was accepted officially.9 Is there a Safe Limit for Nitrate? The two previous sections show that any safe limit for nitrate needs to be related to methaemoglobinaemia rather than stomach cancer. 'Blue-baby' cases in the USA were associated2 with nitrate concentrations ranging from 283 to 1200 g m^{-3}. The only fatal case in the UK arose from a nitrate concentration of 200 g m^{-3}. The non-fatal case in the same report involved bacterially polluted water with 95 g m^{-3} of nitrate. There is no obvious safe level, but in evidence to the House of Lords1 three august bodies, The Medical Research Council, The Institute of Biology, and the Institute of Cancer Research, stated that the majority of cases have occurred when the water contained more than 100 g m^{-3} of nitrate. The limit of 50 g m^{-3} imposed by the European Commission thus has a safety factor of two. This safety factor is a mixed blessing. To some it represents prudence, but to farmers and water suppliers it is a problem. A limit of 100 g m^{-3} for water draining from agricultural land would be quite easy to achieve, but 50 g m^{-3} is far more difficult,1 particularly in the drier areas of the UK and when there is a substantial input of nitrogen to the soil from the atmosphere.

Mortality from cancer of the stomach and other causes between 1 January 1946 and 28 February 1981 among 1327 male workers in a factory making ammonium nitrate fertilizer. Deaths observed compared with deaths expected from statistics for workers in comparable jobs in the locality. Statistics for heavily and less heavily exposed workers combined

	Number of deaths	
Disease†	*Observed*	*Expected*
Stomach cancer	12	12.06
All cancers	91	86.83
Respiratory diseases‡	36	51.04
Heart disease‡	92	113.72
All causes	304	368.11

Nitrate as an Environmental

Problem It is not only land plants that need nitrogen for growth. Plants growing in water respond to extra nitrogen like crop plants, but their extra growth is not welcome. Increased nitrate concentrations in rivers and lakes may encourage reeds to grow to excess, narrowing waterways and possibly overloading and damaging banks. Underwater plants also proliferate, so that anglers lose tackle, the propellers of boats get fouled and conduits for water supply become clogged and machinery damaged. Large water plants, though a nuisance, are not the main environmental problem. Algae are very small single-celled plants that grow on a variety of surfaces, including that of water. They are not noticeable on water until they grow to excess and form the 'blooms'—possibly better described as scum—that are now seen to a worrying extent on our rivers and lakes. These blooms look messy when they grow, but they become a far bigger problem when they die. The bacteria that decompose them use oxygen when they do so and thus deprive fish and other desirable organisms, which may die. The whole ecological balance of the river or lake may change, usually to the detriment of species that we would like to see there. This process, known as 'eutrophication', is limited by phosphate rather than nitrate in fresh water, and it is discussed in detail elsewhere in this issue. Algae can also grow to excess in the sea, but these marine blooms seem to be stimulated more by nitrate than by phosphate. Algal blooms took on a new significance a few years ago when it was discovered that some species were toxic to humans and dogs, a problem that is again discussed in more detail elsewhere in this issue. The indirect effects of nitrate in water proved in this instance to be a far more tangible health hazard than the direct effects.

REFERENCES

1. Arah, J.R.M. and A.J.A. Vinten, 1995. Simplified models of anoxia and denitrification in aggregated and simple-structured soils. Eur. Journal of Soil Science 46, 507-517.
2. Groenendijk, P., L.V. Renaud and J. Roelsma, 2005. Prediction of Nitrogen and Phosphorus leaching to groundwater and surface waters; Process descriptions of the Animo4.0 model. Wageningen, Alterra–Report 983, 114 pp.
3. Heinen M., 2006. Simplified denitrification models: Overview and properties. Geoderma 133, 444-463.
4. Hendriks, R.F.A., Wolleswinkel, R. & Akker, J.J.H. van den, 2007. Predicting soil subsidence and greenhouse gas emission in peat soils depending on water management with the SWAP-ANIMO model. In B. Robroek, et al., (Eds.), Proceedings of the First International Symposium on Carbon in Peatlands, Wageningen, The Netherlands, 15 - 18 April 2007 pp. 113-114.
5. Kersebaum, K.C., J-M. Hecker, W. Mirschel and M. Wegehenkel, 2007. Modelling water and nutrient dynamics in soil–crop systems. In: Kersebaum, K.C. et al. (Eds). Proceedings of the workshop on "Modelling water and nutrient dynamics in soil–crop systems" held on 14–16 June 2004 in Müncheberg, Germany. Springer, The Netherlands. pp 1-17.
6. Lewis D.R. and M. B. McGechan, 2002. A Review of Field Scale Phosphorus Dynamics Models. Biosystems Engineering 82, 359–380.
7. McGechan M. B. and D. R. Lewis, 2002. Sorption of Phosphorus by Soil, Part 1: Principles, Equations and Models. Biosystems Engineering 82, 1–24.
8. McGechan M.B., 2002. Sorption of Phosphorus by Soil, Part 2: Measurement Methods, Results and Model Parameter Values. Biosystems Engineering 82, 115–130.
9. McGechan, M.B. and L. Wu, 2001. A Review of Carbon and Nitrogen Processes in European Soil Nitrogen Dynamics Models. In: Shaffer, M.J., L. Ma and S. Hansen (Eds.), Modelling Carbon and Nitrogen Dynamics for Soil Management. Lewis Publishers, Boca Raton. pp. 103-167.
10. Moreels E., S. De Neve, G. Hofman and M. Van Meirvenne, 2003. Simulating nitrate leaching in bare fallow soils: a model comparison. Nutr. Cycling Agroecosyst 67, 137–144.

11. Renaud, L.V., J. Roelsma and P. Groenendijk, 2005. User's guide of the ANIMO4.0 nutrient leaching model. Wageningen, Alterra–Report 224, 184 pp

12. Reiniger, P, J. Hutson, H. Jansen, J. Kragt, H. Piehler, M. Swarts, H. Vereecken, 1990. Evaluation and testing of models describing nitrogen transport in soil: a European project. In: Transaction of 14th ICSS, Kyoto, Japan, August 12-18, 1990. Volume I: 56-61.

13. Rijtema, P.E., P. Groenendijk and J.G. Kroes, 1999. Environmental impact of land use in rural regions. The Development, Validation and Application of Model Tools for Management and Policy Analysis. Series on Environmental Science and Management vol. 1, Imperial College Press, Imperial College, London, UK, p. 321.

14. Shibu, M . , P. Leffelaar , H. Van Keulen and P. Aggarwal, 2006. Quantitative description of soil organic matter dynamics—A review of approaches with reference to rice-based cropping systems . Geoderma 137, 1 – 18.

15. Schoumans, O.F. and M. Silgram (eds.), 2003. Review and Literature Evaluation of Quantification Tools for the Assessment of Nutrient Losses at Catchment Scale. EUROHARP report 1-2003, NIVA report SNO 4739-2003, Oslo, Norway, p. 120.

16. Vereecken, H., E.J. Jansen, M.J.D. Hack-ten Broecke, M. Swerts, R. Engelke, S. Fabrewitz & S. Hansen, 1991. Comparison of simulation results of five nitrogen models using different datasets. In: CEC, 1991. Final report of the project: Nitrate in soils (Sect. 8.2).

17. Willigen, P. de, 1991. Nitrogen turnover in the soil-crop system; comparison of fourteen simulation models. Fertilizer Research 27, 141-149.

18. Willigen, P. de., O. Oenema and W. de Vries, 2007. Modelling Nitrogen and Phosphorus Cycling in Agricultural Systems at Field and Regional Scales. In: Marschner, P. and Z. Rengel (Eds.) Nutrient Cycling in Terrestrial Ecosystems. Book Series Soil Biology Vol. 10, Springer. pp 361-390.

19. Wu, L. and M. B. McGechan, 1998. A Review of Carbon and Nitrogen Processes in Four Soil Nitrogen Dynamics Models. J. Agric. Engng Res. 69, 279-305.

20. Blanco Gómez, P., 2007. Simulation of water flow and DOC transport in upland peat soils with different land covers using SWAP and ANIMO models. MSc Thesis, School Of Applied Sciences, Cranfield University, UK.

21. Bondarik, I. G. and V.A. Koryagin, 2004. Use of mathematical models SWAP, ANIMO, WOFOST, in farm production scenario analysis for reclaimed agro-landscapes in Kaliningrad and Moscow regions. Proceedings of ICID Interregional Conference on food production and water: social and economic issues of irrigation and drainage, Moscow, Russia, 5-11 September 2004. ICID.

22. Dawson, Q. L. 2006. Low-lying agricultural peatland sustainability under managed water regimes. Ph.D. Thesis, National Soil Resources Institute, Cranfield University, UK.

23. Hack-ten Broeke M.J.D., W.J.M. de Groot and J.P. Dijkstra, 1996. Impact of excreted nitrogen by grazing cattle on nitrate leaching. Soil Use and Management 12 190-198.

24. Hack-ten Broeke M.J.D. and A.H.J. van der Putten, 1997. Nitrate leaching affected by management options with respect to urine-affected areas and groundwater levels for grazed grassland. Agriculture, Ecosystems & Environment 66: 197-210.
25. Hack-ten Broeke M.J.D. and W.J.M. de Groot, 1998. Evaluation of nitrate leaching risk at site and farm level. Nutrient Cycling in Agroecosystems 50: 271–276.
26. Hendriks, R.F.A., K. Oostindie and P. Hamminga, 1999. Simulation of bromide tracer and nitrogen transport in a cracked clay soil with the FLOCR/ANIMO model combination. Journal of Hydrology 215, 94–115.
27. Kroes, J.G. and J. Roelsma, 2007. Simulation of water and nitrogen flows on field scale: application of the SWAP-ANIMO model for the Müncheberg data set. In: Kersebaum, K. Ch., et al., (Eds.), Modelling Water and Nutrient Dynamics in Soil-Crop Systems. Springer. pp. 111-128.
28. Maria Elena Ruiz / Jose Miguel de la Paz / Carlos Ramos / Hanoi Medina SISTEMA SWAP-ANIMO PARA ESTIMAR LOS BALANCES DE AGUA Y NITRATOS EN UNA REGIÓN DE VALENCIA, ESPAÑA. Revista Ciencias Técnicas Agropecuarias, año/vol. 17, número 001 Universidad Agraria de La Habana. La Habana, Cuba.
29. Marinov, D., Querner and J. Roelsma, 2005. Simulation of water flow and nitrogen transport for a Bulgarian experimental plot using SWAP and ANIMO models. Journal of Contaminant Hydrology 77 145– 164.
30. Mioduszewski, W., M. Fic, A. Slesicka, A. Zdanowicz, W. Walther, M. Paetsch, F. Reinstorf, D. Weller, S. Diankov, G. Velovsky, S. Radoslavov, W. Marinov, O. Nicheva, E.P. Querner and J. Roelsma, 2005. Development of tools needed for an impact analysis for groundwater quality due to changing of agricultural soil use. In: Nitrates in groundwater. European meeting of the International Association of Hydrogeologists; Wisla (Poland). Balkema, Leiden. pp. 123 - 128.
31. Rijtema, P.E. and J.G. Kroes, 1991. Some results of nitrogen simulations with the model ANIMO. Fertilizer Research 27, 189-198.
32. Salm, C. van der and O.F. Schoumans, 2000. Phosphate losses on four grassland plots used for dairy farming; Measured phosphate losses and calibration of the model ANIMO. Alterra, Wageningen, The Netherlands. Report 083.
33. Eugenio Carrillo Ávila. Sobre el uso del método inverso para la caracterización hidrodinámica de los suelos
34. Ten Berge, H.F.M. (Ed.), 2002. A review of potential indicators for nitrate loss from cropping and farming systems in the Netherlands. Plant Research International, Wageningen. The Netherlands. Report 31.
35. Vinten, A.J.A., 1999. Predicting Nitrate Leaching from Drained Arable Soils Derived from Glacial Till. J. Environ. Qual. 28, 988-996.
36. Vinten, A.J.A., B.C. Ball, M.F. O'Sullivan and J.K. Henshall, 2003. The effects of cultivation method, fertilizer input and previous sward type on organic C and N storage and gaseous losses under spring and winter barley following

long-term leys. J. of Agr. Sci., 139, 231-243.

37. Wolf, J., M.J.D. Hack-ten Broeke and R. Rötter, 2005. Simulation of nitrogen leaching in sandy soils in The Netherlands with the ANIMO model and the integrated modelling system STONE. Agriculture, Ecosystems and Environment 105, 523–540.
38. Boogaard, H.L., Kroes, J.G. 1998. Leaching of nitrogen and phosphorus from rural areas to surface waters in the Netherlands. Nutr. Cycling Agroecosyst. 50, 321-324.
39. Bijlsma R, P. Groenendijk, P. Boers and M. Blind, 2006. Uncertainty assessment on the nutrient concentration in the Regge catchment, Vecht River basin, the Netherlands. HarmoniRiB report D7.4
40. Groenendijk, P. and P.C.M. Boers, 1999. Surface water pollution from diffuse Agricultural sources at a regional scale. In: Heathwaite, L., Ed. Impact of Land-Use Change on Nutrient Loads from Diffuse Sources. Wallingford (UK), IAHS Publ. No. 257, pp. 235-244.
41. Groenendijk, P., R. Bijlsma, D.J.J. Walvoort & L.V. Renaud, 2008. Analysis of nutrient losses and parameter uncertainties in soils and surface water systems at the catchment scale. In: Abesser, C., T. Wagener & G. Nuetzmann, Groundwater – Surface Water Interaction: Process Understanding, Conceptualization and Modelling, 158 – 163, IAHS Press, Wallingford, UK.
42. Hack-ten Broeke, M.J.D. A.G.T. Schut and J. Bouma, 1999. Effects on nitrate leaching and yield potential of implementing newly developed sustainable land use systems for dairy farming on sandy soils in the Netherlands. Geoderma 91,217–235.
43. Kroes, J.G., P.C.M. Boers, H.L. Boogaard, E.F.W. Ruijgh and J.A.P.H. Vermulst, 1997. Impact of manure policies on leaching of nitrogen and phosphorus from rural areas to Dutch ground and surface waters: methodology and application on a national scale. In: Brouwer, F.M. and W. Kleinhans (eds.), 1997. The implementation of nitrate policies in Europe: Processes of change in environmental policy and agriculture, Kiel, published by: Wissenschaftsverlag Vauk, Kiel, Germany, ISBN 3-8175-0262-1
44. Kroes, J.[G.], J. Beldman, H. te Beest, D. Boland & T. Vellinga, 2002: Conflicting interests in a Dutch agricultural dairy farming system. In: J.[H.A.M.] Steenvoorden, F. Claessen & J. Willems (eds.), Agricultural effects on ground and surface waters: research at the edge of science and society. Wallingford (UK), IAHS, 2002. IAHS Publ. 273, pp. 41-48.
45. Kroes, J.G. and P. Groenendijk, 2003. Nutrient emissions from soil to surface waters: integrated modelling at a regional scale. Geophysical Research Abstracts, Vol. 5, 14386.
46. Ollesch G., I. Kistner, Y. Sukhanovski and M. Rode, 2005. Dynamic and modelling of sediment associated nutrients in a low mountain environment. In: Horowitz, A. J. and D.E. Walling. Sediment budgets 2. IAHS Press, Wallingford, Vol. 292, pp. 171-178.

47. Ollesch G., I. Kistner, R. Meissner and K.-E. Lindenschmidt, 2006. Modelling of snowmelt erosion and sediment yield in a small low-mountain catchment in Germany. Catena 68 161 – 176.

48. Sonneveld, M.P.W. and J. Bouma, 2003. Methodological considerations for nitrogen policies in the Netherlands including a new role for research. Env. Sci & Policy 6 501–511.

49. Sonneveld, M.P.W. and J. Bouma, 2003. Effects of combinations of land use history and nitrogen application on nitrate concentration in the groundwater. NJAS 51, 135-146.

50. Steenvoorden J.H.A.M., C.W.J. Roest, and P.C.M. Boers 1997. Simulation of nutrient losses to groundwater and surface waters in The Netherlands. In: Webb, B. Ed. Freshwater Contamination. Wallingford (UK), IAHS Publ. No. 243, pp. 323-332.

51. Wolf J., A.H.W. Beusen, P. Groenendijk, T. Kroon, R.Rötter and H. van Zeijts, 2003. The integrated modeling system STONE for calculating nutrient emissions from agriculture in the Netherlands. Environmental Modelling & Software 18, 597–617.

52. Rijtema, P.E. and Kroes, J.G, 1990. Nitrogen modelling on a regional scale. Proceedings of the NATO ARW workshop "Nitrate-contamination: Exposure, Consequence and Control", Lincoln, Nebraska, sept.1990.

Index